The Calculus Dire

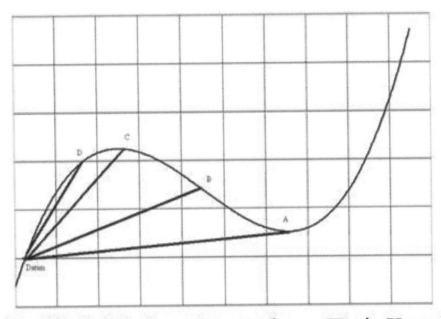

An Intuitively Obvious Approach to a Basic Knowledge of the Calculus for the Casual Observer

The Calculus Direct
An Intuitively Obvious Approach to a Basic Knowledge of the Calculus for the Casual Observer
By John Weiss
2009

The Calculus Direct

Preface:

The idea for this book was formulated from my experience as an adjunct mathematics instructor at Northwest Vista community college. I became involved with the developmental mathematics in 2006 and was appalled to learn that a student without strong mathematical skills who hoped to attain a technological degree had to complete 7 semesters of mathematics to advance through calculus. This is worrisome because the Calculus is the most important subject in modern education. Physics, engineering, biology, rocket science, and the electrical and computer systems that govern our lives derive all of their abilities from the calculus and not the algebras. With mastery of the calculus entire subjects are laid bare for the student to easily grasp. Because of this, developing a competent student in this area is critical in our modern world. Unfortunately the calculus is placed at the top of a mountain of mediocre material that systematically bores and frustrates the average prospective student to the point of hating mathematics. Although this 'mediocre material' is useful in some respects it is not critical to the modern mind. Just as Latin has fallen from a required part of education in our modern times so must most of the 'curious' but altogether useless algebraic puzzles that we constantly harass the modern student with. I make the statement that if we are ever to raise the average student to an appreciable technical understanding we must shun our antiquated embrace of menial algebraic puzzles and instead open the doors to modern mathematical ponderings. Although a thorough understanding of algebra is required, one does not need to master algebra before calculus can be studied. It is on this principal that I write this book. My goal is to take a willing student who is at the lowest level of mathematics and achieve a workable understanding of the calculus. In essence we will start with addition, and taking no previous mathematical knowledge for granted, follow the most efficient path to integration. If I can show, as I believe I have, that a basic understanding of Calculus can be found through one concise book then my point will be proven.

This book is not the end all of mathematical education, it is a starting point. I have left certain recommendations open and in fact explicitly state topics for further reading. I hope the student finds this book illuminating and proceeds to further exploration of the natural world and the seemingly magical connection with mathematics. It fills me with wonder that mere thought (mathematics) could actually mirror and even predict this ether surrounding us that we refer to as reality. Maybe, hopefully, this will pique your interest as well.

The intended audience for this book is the non-traditional student who is entering college without a firm understanding in mathematics and wishes for supplemental material. It is also intended for the uninformed yet curious souls tinkering in their garage and trying to make sense of a world that has left them behind technically. It is my sincerest belief that these types of people have been and always will be the backbone of American technological innovations

As an aside for those wayward mathematical purists, this is not a completely rigorous and solid approach. There may be sections that you disagree with and dislike as far as the mathematical approach is concerned, but keep in mind that the purpose is to open doors for nontraditional as well as weak mathematical readers. Some leniency in rigor is required for these type of students.

As an aside to those readers who hope this to be a magical book… unfortunately you still need to put in the effort to understand the calculus. I have written what I believe to be the route with the least amount of work, but I do not know of a magic bullet that automatically produces understanding by merely holding the book and eating cake. Some work on the readers part is expected, even demanded, for the reader to understand everything. This does not mean it is hard or discouraging, but some effort is needed.

Chapter 1.1 Addition

The numberline is the basis of all mathematics and we will begin our journey here. Our first lesson is how to create the numberline. We do this by starting with a line and then choosing a position on the line as a datum point. A datum point is an origin on which all other values are based upon. At this home point we list our home number - zero. Now, choose some arbitrary measurement and call this a unit measurement. We then measure one of these unit measurements to the right of zero (we chose the right of zero because we read from left to right) and we find a position which we naturally label 1 to signify one unit from zero. Measure another unit, this time starting at our labeled point 1, and we find point 2 signifying (no prizes) two units away from zero. Repeat this process repeatedly until we reach infinity. Infinity is a number that is larger than any number you can think of. We like to say that it is somewhere in the distance far - far to the right of zero and leave it at that because people begin to go crazy every so often when they study infinity too much. I'm not joking, institutionalized. As such we will just define infinity as the furthest point to the right and continue to more pressing matters. Infinity, in our case, looks to be larger than seven and this should feel correct.

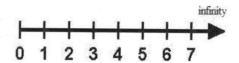

Let's take that same unit measurement we used to create the right side of the numberline and measure that same distance to the left of zero. But what will we label this point on the numberline? Since our values continue to become smaller as we move from right to left we will need a number that continues this pattern. But what number is less than zero?

In tangible life we cannot have anything less than zero. However, and this is a very important mathematical lesson, we are interested in the idea and not the reality. It doesn't matter to us that we can't hold a number less than zero in our hand, only that we can describe it, and we describe it as -1 (pronounced negative one). Measuring another unit value from -1 and traveling further to the left we find -2, repeating we get -3, -4, and so on until we find negative infinity. Negative infinity is considered the number furthest left of every other number and the least number... if that makes sense. But we won't study infinity until you're older.

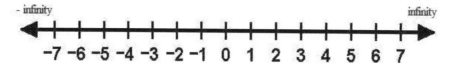

This is known as the integer numberline. An integer means any whole number positive or negative including zero. Numbers, as we have seen, are a symbolic form for writing value. Each corresponds to a particular group of units. However, to truly appreciate math, we need to create a new understanding of numbers and the numberline allows us to define a way of visualizing this. For instance, when we say the number 5, the number of fingers on a hand, we are indicating a length of five units measured to the right of zero. When we say -10 we indicate a group of ten units to the left of zero. To further examine this new way of thinking the student will be introduced to a vector.

A vector has both magnitude (distance) and orientation (direction). The scalar of ten only consists of a group of units. However, the vector of 10 situated on the number line consists of both the distance from 0 to 10 and also a direction traveling to the right. The vector negative ten (-10) consists of the distance from 0 to 10 also, but has a direction traveling to the left. We can think of these vectors as entities onto themselves. They do not necessarily have to start at 0 and end at their respective point assigned to them. It can easily be argued that if we started and stopped between two arbitrary numbers, say 20 and 25, we would have five units between these numbers. What about the vector from positive 8 to 4? Well, the direction from 8 to 4 is to the left, thus negative, and the amount of units between these two points is four units, our final answer is -4. We will utilize this type of thinking shortly to begin our adventure into algebra.

Another critical observation brought about by the numberline is that we have symmetry about 0. Symmetry means that the mirror image is reflected over a particular point. Notice how 5 and -5 have the exact same magnitude (distance) but different directions relative to zero. We call these two vectors opposite vector pairs. But it isn't just with five. All of the numbers share this quality with their negative counterpart.

Mathematics main goal is to describe the relationship between the points on the numberline by internationally recognized symbols. The most basic symbol involving the numberline is the equal sign (=). An equal sign sets a precedent between two arguments stating that the value of one idea is the exact same as the value of the second idea. In regards to the numberline, this means that both ideas occupy the exact same spot on the numberline. Thus when I say 1=1 I am stating that a symbol is equal to itself and therefore occupies the exact same spot on the numberline. No kidding right? Just wait it gets more interesting. The next unknown will be what happens if I combine both values together? To do this we need to introduce the addition sign (+). When we encounter the addition sign (+) it alerts us to combine two groups into one. We now move into the question; what happens if we add two vectors of one together? In mathematical symbols:

$$1 + 1 = ?$$

Most of the students have simply rolled their eyes and said 2. But you are doing arithmetic, and the problem with arithmetic is that it is problem specific. We wish to train you in algebra. Algebra deals in the generalized application of logic rather than the specific instances of certain problems. The point of this book is to focus on the logic of mathematics.

So the question is presented again, what is the collection of two arbitrary values? It turns out our numberline is very instructive in explaining our first problem presented to mathematicians. But to do this we need to deconstruct our arithmetic numberline above by stripping away every single number except for our datum point zero.

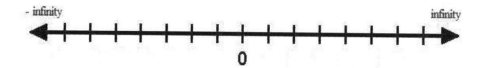

Much better, now we will populate our new algebraic numberline with an idea. Consider a vector symbolized by the variable b. A variable is a tricky subject. Basically it simply holds the position of a quantity that is absolute but at this point is a deformable unknown symbolized by a letter. Finding the unknown value of the variable will be the main effort of algebra. The variable could symbolize a group of horses, or marbles, both, or what-have-you. In all honesty we don't care. Our only concern is that the quantity can be found through the use of applied algebra. It is important to pull yourself away from concrete objects in mathematics and move towards malleable ideas.

Rephrasing our argument we have a vector **b** that symbolizes our idea, it moves to the right and has some measurement. We wish to add another vector, say **c**, that symbolizes another idea and moves to the right also.

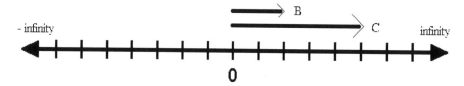

It can easily be seen that **b** units to the right combined with another **c** units to the right culminates in both distances stacked towards their particular direction into another vector that we will call **x**

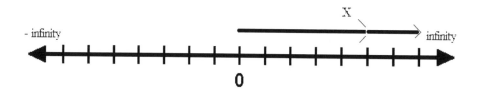

$$b+c = x$$

Thus to add our two vectors we simply combine the vectors head to tail into a new unit value. Furthermore, we can see that it doesn't matter which vector we choose to represent first. To help prove this to yourself visualize a piece of string cut into a small and larger section. If we first lay down the larger piece and then attach the smaller to it, wouldn't we have the length of the original piece of string? Now visualize that you first place the smaller string down and then attach the larger piece. Wouldn't we again find our original length? The same idea works with numbers and we can interchange addition freely. This ability to interchange numbers when adding without affecting the final answer is known as the commutative property of addition.

Commutative property of addition:

$$\text{Eq 1.1.1} \qquad b+c=c+b$$

Above I defined addition as combining the two vectors head to tail based upon zero. This wording might seem longwinded but it had a purpose. Consider another vector d, which moves from right to left.

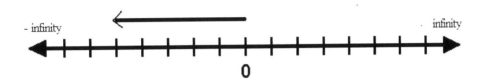

We know that d is a negative number because of our definition of negative numbers. But we wish to add d to our original vector b anyway. We will accomplish this addition exactly as before, we will combine the two vectors head to tail centered at zero. This will culminate in a new vector which we will label y.

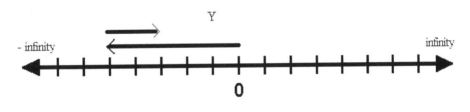

$$b+d=y$$

Thus the addition of negative numbers will give us no great trouble because the process is the same as adding positive numbers.

The equals sign
An equal sign is a powerful sign that defines an exact convergence between two ideas. The equal sign, already introduced, is so powerful because once two separate ideas are equated they may be substituted freely. We can call something equal if and only if the two ideas terminate at the exact same spot on the number line. Thus:

$$1=1$$

We know this because both vectors are equal to one unit moving to the right. We can also show that.

$$3+1=6-2$$

Because if we were to plot these on the numberline we would see that each vector ends at 4 exactly and thus equal to each other.

But again we don't care about numbers, we wish to look at variables. For instance recall that positive numbers are to the right of zero and negative numbers are to the left of zero. Now consider our two vectors c and d measured against each other. We can see that they are the same size in magnitude but differ in direction.

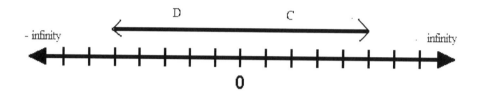

Clearly we can tell that they are mirror images of each other, which was our definition of opposite vector pairs as argued above. Thus c is the negative value of d. In mathematical terms:

$$-c = d \qquad \text{and likewise} \qquad c = -d$$

Plotted on the numberline

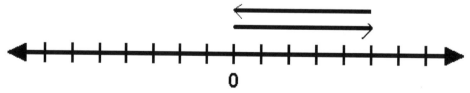

This solidifies our opposite pair theories showing that both vectors are mutually negative to each other. With this in mind we can approach an interesting observation. What would happen if we were to add two opposite vector pairs together? Following our understanding of addition.

Notice how our vector c moves some distance, but that d erases any progress c might have made. This can clearly be explained if you substituted the value of d for its equivalent representation in c (namely $-c = d$)

The addition of opposite pairs also sets up two corollaries (connected results). First we know right off what would happen if we were to add any pair of opposite vectors.

$$\text{Eq 1.1.2} \qquad c+d=c+(-c)=(-d)+d=0$$

Because we have created an exact convergence between two ideas we can substitute them whenever and wherever we feel the need. No matter how we slice it, we will always find 0 based upon our operation of addition. With this knowledge we can quickly see:

$$b+c+d=b+c-c=b+0=b$$

With no uncertain terms we come to something that seems trivial, but is critical. If we add a number to an original quantity and then subtract that same number we will be left with our original starting quantity. In other words, any number plus zero is our starting number.

$$\text{Eq 1.1.3} \qquad\qquad b + 0 = b$$

Subtraction

In mathematics there are operations, and then counter operations. For addition the counter operation is subtraction. Addition is concerned with combining arbitrary vectors to our original quantity. The opposite of this, subtraction, is interested in removing arbitrary values from our original quantity. We will be using the same set of vectors we used to discuss addition, meaning that c and d are opposite pairs. Now consider that we wish to subtract the vector c from our principal vector b.

$$b-c=?$$

Since we have already defined (-c) as equal to a positive d. We can substitute this value into our equation.

$$b-c=b+d$$

We have just found a very important simplification in mathematics. Since d is the opposite of c, when we subtract we are actually adding the opposite. Subtraction is now defined by addition, and any logical argument that works for addition has to work for the special case of subtraction.

$$\text{Eq 1.1.4} \qquad \begin{aligned} b-c&=b+d=b+(-c)\\ b-c&=b+(-c) \end{aligned}$$

Now we can interchange the order of these vectors because subtraction is defined in addition, thus any rules of addition apply to subtraction as well:

$$b+(-c)=(-c)+b$$

One very important note! Notice how the negative sign is part of the variable and moves with the variable. Many students forget to move the negative sign along with the variable.

This type of logical simplification of placing subtraction within the umbrella of addition is exactly what we wish to provide to the student in this book. Why this is so important is because we have effetely cut our things to remember by half. Everything that works for addition will work equally well for subtraction. Numerical examples follow below:

Numerical Examples:

1) -2+5
- To complete this sum we take our two vectors and place them together head to tail. Where the second head ends is our answer.

$$-2+5=3$$

2) -2-5
- First thing we will notice is that when we are subtracting a number we are actually adding the opposite pair.
- The opposite pair of 5 is a negative 5
- Rewriting our equation with this interpretation in mind.

$$-2-5=-2+(-5)$$

- Then applying our understanding of two vectors added together.

$$-2+(-5)=-7$$

3) $5+(-2)$
- This is a simple addition of two vectors.
- Recalling the commutative property of addition we should find the same answer as the 1st example.
- Following our established procedure we find:

$$5+(-2)=3$$

- Notice the answer is exactly the same as in example 1.

4) $5-(-2)$
- Again, if we are subtracting then we really are adding the opposite pair
- The opposite pair of -2 is a positive 2.

$$5+2$$

- Combining our vectors through the usual way we find.

$$5+2=7$$

- Notice that we get a different answer than example 2. One is negative while the other is positive, and in mathematics, that means the entire world.

Mathematics consists of processes independent of the number. You must remove the number from your thinking and instead dwell on the idea and process of the underlying logic. The faster you do this, the quicker math will begin to make sense to you. Then maybe your life, but defiantly your grade, will get better.

Chapter 1.2 - Multiplication and division

Consider some aggregating periodic process. This could be a salaried paycheck, the number of classes you have in a week, or the number of miles you drive to work each day. No matter the reason, we simply have a repeated accumulation of a specific number. Again it is important for the student to train the mind off of tangible circumstances and instead focus on malleable ideas. Let's define one of these repeating groups as b. In terms of our newly accepted vectoral representation of addition, our process begins with a single b and then adds another vector b.

$b+b =$

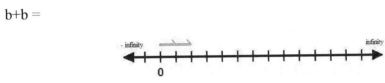

We can see that the distance b+b covers is twice the distance of the original b. lets add another b to the mix.

$b+b+b =$

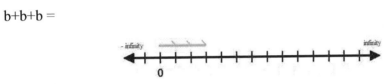

now we have thrice the distance that the original b held. Adding more and more.

$b+b+b+b =$

$b+b+b+b+b =$

$b+b+b+b+b+b =$

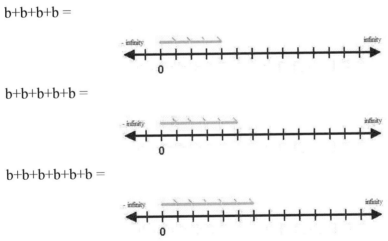

This quickly becomes tedious. Our hand starts to get tired, plus the b's begin to intermingle to our eyes. Does the last line have 7 b's or 8? It is hard for me to tell. Mathematicians are inherently lazy people. We like things to be short and easy, and if it's not we strive to find another way around. Laziness is the mathematician's greatest virtue and drives the wheels of progress.

With the complaint that adding large amounts of the same number is tedious, we develop a short hand way of representing the same idea. Let's focus on the amount of times we add. If we add 1 b to another b, then we should have 2 b's all together.

$$b+b= 2b$$

Adding another b to the mix gives us 3 b's

$$b+b+b=3b$$

Continuing and continuing....

$$b+b+b+b=4b$$
$$b+b+b+b+b=5b$$
$$b+b+b+b+b+b=6b$$

Culminating in the observation:

Eq 1.2.1 $b+b+b+b+b+b+...+b+b=nb$

The ellipses in the equation are a mathematicians way of telling us to continue the 'obvious' pattern. Equation 1.2.1 states that if we add a number of b's together (we call this n times) we will have n groups of addition. n can be any number, and the right hand side of equation 1.2.1 is the mathematical representation of n groups of addition containing the number b. A numerical example is provided at the end of the chapter to further clarify.

This short hand developed above is called multiplication. It allows us to write very large additives in a mere matter of seconds. It also allows us to know exactly how many times we are asked to add a specific number without having to stop and count every time we encounter a new math problem.

One note, we no longer use the sign × to denote multiplication. We still use this sign, but its use is reserved for 3-dimensional multiplication. If you're interested in this there are many books on the subject. In fact, 3-d multiplication is one of my favorite subjects. The sign for multiplication in algebra is no sign at all. That is to say we simply write two symbols next to each other, nb for example, to denote that we want to multiply n and b together. When we are dealing with pure numbers we use parenthesis. 2(4) then means 2 multiplied by 4, or 2 groups of 4 added together.

Another interesting consideration dealing with multiplication deals with multiplying by one. Consider below:

$$1a =?$$

So what is 1 group of a worth? Well obviously it is worth an a! It is defined as one group of that particular thing added together. From this idea it is seen that anything multiplied by 1 is itself.

$$1a = a$$

What about zero? What if we had zero groups of a added together?

$$0a =?$$

Don't jump to conclusions. We will solve this little conundrum later after a few more explorations.

Division

In mathematics we always have a manipulation and the counter manipulation. For addition, it was subtraction. For multiplication, it is division. Since multiplication combines equal groups to create a new idea, division separates our idea into equal groups. Division asks if there is some quantity, how much would one group be worth if we separate this quantity into a specified amount of equal groups. Consider an original quantity d, shown below.

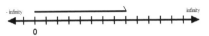

We want to find the exact amount we would need to separate this quantity into 2 equal groups. It can quickly be seen that we would have 2 groups ½ the size of d.

$$\frac{d}{2}$$

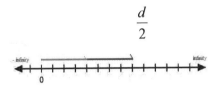

Let's explore what happens when we increase our division to 3, 4, then n.

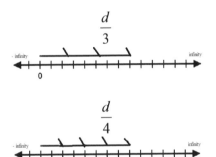

It can be seen that the larger we make the total amount of groups the smaller the group itself becomes. So it is easy to see that d divided by n gives us n groups an nth the size of the original d. In symbolic form this is:

$$\frac{d}{n}$$

This fractional form of division is so useful that we don't use the division sign much anymore. In fact most division you will encounter will be notated as a fraction. There are only a few exceptions.

One important special case of division is how large a single group will be when we divide our vector d into d separate groups. Think if we had any specific number of units, cookies perhaps, and we had exactly the same number of groups to distribute those units to, friends maybe. Let's imagine that we ate too much and have an assortment of cookies that will go to waste. A group of friends arrive and we must ask ourselves how many cookies would each friend get, or how many units would each group get, if both the amount of cookies and the amount of friends were equal. Well it would be 1 unit. This is known as the divisional property of unity.

$$\text{Eq 1.2.2} \qquad \frac{d}{d} = 1$$

An important result we will utilize often.

A second important special case of division rests upon dividing a number by zero. The argument for why this is impossible algebraically is a little tedious. I have found the best thought experiment is the cookies that need to be distributed to friends. How many cookies could you give to zero friends? Well you can't just throw the cookies away, you need to give them to someone. You can't eat them because you're full... what could we do? The algebra runs into the same problem when it divides by zero. We say that there is no solution. This inability to divide by zero is a major downfall in algebra.

Eq 1.2.3

$$\frac{d}{0} = \text{no solution}$$

Division defined by multiplication

Recall multiplication is defined as the grouping of addition. So what if instead of two groups we wanted to describe only half a group? Using our multiplication notation we want one half of a group of d, or:

$$\left(\frac{1}{2}\right)d$$

You might notice that this subject was encountered though our division arguments above in which we defined another way to represent a half group of d. That's right d divided by two. So we can see that d multiplied by 2 is a different interpretation of the exact same idea of d divided by two. We can generalize this argument by replacing n with 2 as below.

$$\text{Eq 1.2.4} \qquad \frac{d}{n} = d\left(\frac{1}{n}\right)$$

Just as subtraction is defined by addition, division is defined by multiplication. We now only have two main mathematical manipulations to remember so far, addition and multiplication. We have effectively cut our things to remember by half!

Important multiplication considerations:

A funny thing happens in arithmetic. n groups of b units is the exact same amount as b groups of n units. In symbolic notation:

$$\text{Eq 1.2.5} \qquad n(b) = b(n)$$

This can be seen by sorting out our groups into a matrix. Recall that 3 means 1+1+1 and that 5 really means 1+1+1+1+1. So if we wanted to have three groups of five we could set this up into a structured formation, shown below.

$$1+1+1+1+1+$$
$$1+1+1+1+1+$$
$$1+1+1+1+1$$

We will interpret this to be 3 rows of 5 columns. But we could also interpret this as 5 columns of 3 rows. It doesn't matter how we interpret it, only that the total number contained in our formation is 15.

$$(3)5 = 15$$
$$(5)3 = 15$$
$$(3)5 = (5)3$$

Notice that this will work for any variables multiplied together, no matter the magnitude. We call this interchangeability the commutative property of multiplication. We can use this property of multiplication to solve our problem of multiplying by zero. Recall that earlier we left a loose end by asking what is zero groups of a? Well, with the commutative property we can turn the question around and ask… what is a amount of groups of zero added together?

$$a(0) = 0 + 0 + 0 + \cdots + 0 = 0$$

Now it is readily apparent that zero multiplied times any finite number is zero. Yes this result was expected and might not have been worth the time in some opinions. However, we shall soon find some very interesting unexpected results and should never take things for granted in mathematics. Notice how multiplying by zero differs from dividing by zero!

Another concern

Turning our attention to repeated subtractions, consider the following argument.

$$-b-b-b-\ldots-b =$$
$$(-b)+(-b)+(-b)+\ldots+(-b) =$$
$$n(-b)$$

This is brought about by our understanding of subtraction (defined by addition). Recognizing our definition of multiplication it is easily seen that a negative number multiplied by a positive number will produce a negative number. Use the numberline to prove it to yourself. It is easily seen on the numberline that repeatedly moving our point to the left a number of times will only further increase our distance from zero more to the left and therefore still be a negative number. A corollary of this argument is a basic definition of negative numbers. Consider n groups of negative 1 added to itself.

Eq 1.2.6 $$n(-1) = (-1) + (-1) + \cdots + (-1) = -n$$

Recall that repeatedly moving to the left of zero one unit was how the negative side of the numberline was constructed. A great result that we will utilize shortly.

What happens when we have a negative number multiplied by a negative number? At first glance this makes no sense. How can we have negative groups of negative numbers? It will turn out to be a positive number, which might not make sense, but remember mathematics doesn't necessarily need a direct human understanding to play its course. Many times we let the math tell us what happens during an unknown process. This is important for us to understand. We do not force the math into doing anything that is illogical. We can only apply logic to the math and watch what comes out. Repeat; do not force the mathematics into anything. Let the mathematical logic guide you! We will first look at another thought process before we move to the idea behind a double negative multiplication.

Multiplication of groups of addition

Consider the following expression:

$$c(d+e)$$

What this expression stands for mathematically is that we wish to add multiple groups of a summation of two terms, d and e being those terms. To see what the multiplication of a general addition is we will set up a simple argument. Consider multiplying the summation by 1. Remember, that's one group of d+e

$$1(d+e)=d+e$$

Easy enough, what if we wanted to multiply by two? Following the definition of multiplication

$$2(d+e)=(d+e)+(d+e)=$$
$$d+d+e+e=2d+2e$$

Recall that by the commutative property of addition (Equation 1.1.1) we can interchange the order of the addition without changing the final value of the answer. Thus we have 2 groups of addition of d and 2 groups of addition of e. Which is simplified into 2 multiplied by d and 2 multiplied by e. Extending our argument to 3

$$3(d+e)=$$
$$(d+e)+(d+e)+(d+e)=$$
$$3d+3e$$

Continuing this argument up to some value c we can see the generalized expression as:

$$c(d+e)=cd+ce$$

This is known as the Distributive property of multiplication over addition. It should be readily apparent that c will distribute to any number of terms inside the parenthesis. In math speak this would appear as…

Eq 1.2.7
$$c(d+e+f+...+z)=$$
$$cd+ce+cf+...+cz$$

The above line simply states that no matter how large the summation group is, the total number of times the group is repeated is specified by c. Therefore there will be c groups of each particular term. A numerical example follows below. But it would serve the student well to try

The Calculus Direct

and understand this concept before moving to the numerical example. Then the numerical examples will serve to reinforce and guide the student rather than act as a crutch. Moving on, suppose we have two groups of addition and we wish to multiply these two groups together. Mathematically they look something like this.

$$(b+c)(d+e)=?$$

We've never multiplied groups together before and we don't want to make a mistake. How should we proceed? Cautiously, we look to a very powerful tool known as the u-substitution method. In the substitution method we simply choose an idea and substitute that idea for u. Observe:

$$\text{Let } u=(b+c)$$

We have made a statement that when combined with our original supposal will transition us quickly from a slippery slope to sure footing. What we will do is use the power of an equals sign. Recall if something is of equal value then we can easily interchange one idea for another. Observe:

$$(b+c)(d+e)=$$
$$u(d+e)=$$
$$du+eu$$

Which we know is true based upon our last argument. The final step to the substitution method is to remember to REsubstitute.

$$du+eu=$$
$$d(b+c)+e(b+c)$$
$$db+dc+eb+ec$$

Basically, when multiplying groups of addition together, every single term inside the first set of parenthesis must be multiplied by every single term within the next set of parenthesis. Always remember to combine like terms, a topic we will cover shortly. This method can be extended for any amount of summation groups

Eq 1.2.8
$$(a+b+...+y)(c+d+...+z)=$$
$$ac+ad+...+az+bc+bd+...+yz$$

Factorial!
Sometimes in mathematics we come across the factorial. We denote the factorial as

$$b!$$

Where b is any integer. The exclamation mark means that we will be multiplying by every single positive whole number between 1 and b. thus if we want to find the factorial of b we would denote it as:

Eq 1.2.9 $\quad b!=(b)(b-1)(b-2)(b-3)...(1)$

One definition that we have to declare is the zero factorial or 0!. This is equal to 1. It is just a definition and there is nothing to understand. As always a numerical example can help to understand the idea and a few are included at the end of the chapter.

Double Negative multiplication

It was propositioned earlier that two negative numbers multiplied together is a positive product. We will look for why this is presently. Any time we multiply two negative numbers we can simplify to a point but then we lack the knowledge to move further. Consider:

$$(-c)(-d)$$
$$(-1)(c)(-1)(d)$$
$$(-1)(-1)(c)(d)$$

Where we have taken our definition of negative numbers and then used the commutative property of addition to rearrange the products. We know what the product of a positive c and positive d is, so the question now lies on what is -1 multiplied by -1? The following argument is one way to shed light on our problem. Let us set the product to some unknown quantity, call it x

$$(-1)(-1) = x$$

Our goal will now be to find a way to define x. In our first step we will take the addition property of zero found in equation 1.1.2:

$$(-1)(-1) + 0 = x$$
$$(-1)(-1) - 1 + 1 = x$$

Recalling that -1 is one group of -1, as shown in equation 1.2.6, we can use the definition of multiplication.

$$(-1)(-1) + (1)(-1) + 1 = 0$$

Reversing the distributive property of multiplication over addition, equation 1.2.7, we can factor out the common factor of negative 1 from both terms on the left hand side.

$$(-1)(-1 + 1) + 1 = x$$

We know that this is correct because if we were to redistribute a negative one to both terms in the brackets, we would have the exact same statement as the equation above. Looking inside our parenthesis we see that we have -1+1=0. Recall anything finite times zero is zero.

$$(-1)(0) + 1 = x$$
$$0 + 1 = x$$
$$1 = x$$

So all along x was equal to a positive 1 as we have just shown. Thus two negative numbers multiplied together will produce a positive number. This may seem counterintuitive but math is logic and the logic of the past argument has no holes, well none that I could see. Thus if the logic holds without contradiction the argument has to be right no matter how ugly, unintelligible, or otherwise unfortunate the result. A Stupid way to live your life right? Well maybe… maybe not.

The Calculus Direct

This is the first unexpected result that careful mathematical analysis reveals to us. More to come later. Finally

$$\text{Eq 1.2.10} \qquad (-1)(-1) = 1$$

Numerical Examples

1.]   (2)(5)

- We can interpret this two different ways, we can see this as five groups of two added together, or we can see this as 2 groups of 5 added together, either way we get

$(2) + (2) + (2) + (2) + (2) = 10$

$(5) + (5) = 10$

- Thus 2 multiplied by 5 is 10.

2.]   (-5)(-2)(-3)

- Here we see three numbers multiplied together. This may seem confusing but we are just looking for groups of groups. So our method will be to find the product of the first multiplication then multiply by the last number
- Carrying out our first multiplication

$(-5)(-2) = 10$
$(-5)(-2)(-3) = (10)(-3)$

- A negative multiplied by a negative is a positive, as per our argument above. After the first product is found we then set about multiplying the second product.
- Remember that a negative multiplied by a positive is a negative.

$(10)(-3) = -30$

Thus $(-5)(-2)(-3) = -30$

3.] $\left(18\right)\left(\dfrac{1}{9}\right)$

- We can interpret this as 18 groups of 1/9th or we can see this as a ninth of 18. either way the implication begets dividing 18 by 9.

$$\left(18\right)\left(\frac{1}{9}\right) = \frac{18}{9} = 2$$

4]   Evaluate 2!
- To find 2! We simply multiply by every single whole number between 2 and 1
- In this case it happens to only be 2 and 1

    $2!=(2)(1)=2$

5]   Evaluate 5!
- To find 5! We multiply by every single whole number between 5 and 1

    $5! = (5)(4)(3)(2)(1)=(20)(3)(2)(1)=(60)(2)(1)=(120)(1)=120$

6]   Evaluate 8!

    $8!=(8)(7)(6)(5)(4)(3)(2)(1)=40320$
- A rather large number. The factorial increases dramatically as seen as the difference between 2! 5! And 8!.

- Remember that $0!=1$

Chapter 1.3 Fractions and prime factorization

I have dedicated a specific section to fractions since many people are weakest in this area. But to understand fractions we first need to understand factors. Factors are defined as the ideas that when multiplied together produce a desired product. For example, consider the number 24. What numbers, when multiplied together create 24? Below I've tabulated all of the whole factors.

$$
\begin{array}{c}
24 \\
\hline
\begin{array}{c|c}
1 & 24 \\
2 & 12 \\
3 & 8 \\
4 & 6 \\
\end{array}
\end{array}
$$

These are the whole numbered factors of 24. Any pair of these, when multiplied together, gives the product of 24. But there is something deeper than these four simple pair. Almost all of these factors can themselves be factored. 12 can be factored to 3 and 4, 6 may be factored to 3 and 2. 2 however is different, it cannot be factored any further by whole numbers. The only pair of whole numbers that 2 can be factored by is 1 and 2. We call numbers like this prime numbers, since they cannot further be reduced cleanly by division. 3 is another prime number as is 5, 7, 11, 13 and 17. The prime numbers don't stop there. In fact there are an infinite amount of prime numbers.

Returning to our example 24, it would seem that there are 4 sets of factors. But let's continue to factor a pair of factors until we reach prime numbers. Starting with 6(4)

Here we see the factors (3)(2)(2)(2). But hold on, let's try factoring 24 with 8(3).

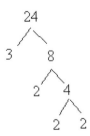

Again we find the exact same set of prime numbers. I make the claim that we don't need to continue this and that all whole number pair of factors of 24 will reduce to this same string of prime factors, which they do. This is known as the Unique Prime Factorization. As the name sounds, the prime factors (3)(2)(2)(2) are unique to 24. No other number in the whole number system has this specific set of prime factors. Verify with the other factors.

Fractions

The first thing for the 'average fraction despiser' to realize is that all numbers are fractions. When I refer to the number five I am actually stating the following:

$$5 = \frac{5}{1}$$

The fraction is also described as a ratio. The ratio describes one idea with regards to, or in the context of, another idea. One easy example would be the speed of a vehicle. We would say something like 5 meters per second. Or

5 meter
second

The fractions parts are called the numerator, the denominator, and the divisor line. They are situated as so:

Numerator
Denominator

With the divisor line separating the upper (numerator) and lower (denominator) parts. The denominator denotes the type of increment we measure while the numerator indicates how many of our increments are present. When we defined 5 above as 5 divided by one, we were describing five whole units. If we were to have (5/2) then we would be stating we have five half units present. If we had (5/3) then that would denote five third units. Finally if we wrote

$$\frac{b}{c}$$

Then we would interpret this as b amount of a cth whole unit. The important thing to notice is that fractions work the same way as whole numbers, but they deal with measurements in increments that aren't the old standby of one. Much like inches and centimeters both can be used to measure a distance. Both provide different numerical answers for the length of the same item but that does not mean that the object changes length due to the use of a different measurement system. Fractions and whole numbers both attempt to measure the distance a number is from zero, and although two different written numbers emerge depending upon our choice of measurement, the actual magnitude of the number is intact… unless we made a mistake of course.

In this way fractions should not be feared. They are the exact same as whole numbers, except fractions attempt to describe a magnitude on a different scale. With this in mind we start our study of fractions in earnest.

Multiplication of fractions

Recall that fractions denote division, and that division is defined within multiplication as so:

$$(a)\left(\frac{1}{b}\right)=\frac{a}{b}$$

We wish to now give meaning to two fractions multiplied together. Something of the form

$$\left(\frac{a}{b}\right)\left(\frac{c}{d}\right)$$

Using first the definition of division within multiplication and then the commutative property of multiplication we will find a quick and easy result.

$$\left((a)\frac{1}{b}\right)\left((c)\frac{1}{d}\right)=$$

$$(a)(c)\left(\frac{1}{b}\right)\left(\frac{1}{d}\right)=$$

$$(ac)\left(\frac{1}{bd}\right)$$

Now, rewriting this multiplication as division we find that to multiply fractions we simply multiply numerator to numerator and denominator to denominator.

Eq 1.3.1 $$\left(\frac{a}{b}\right)\left(\frac{c}{d}\right)=\left(\frac{ac}{bd}\right)$$

Common Term Divided out

In the last section it was shown how to multiply fractions. We will use this property to learn to reduce extraneous factors. Consider a fraction of the form:

$$\frac{ac}{bc}$$

By the definition of multiplication and the commutative property of multiplication we can separate this into two fractions multiplied by each other.

$$\frac{ac}{bc}=\left(\frac{a}{b}\right)\left(\frac{c}{c}\right)=\frac{a}{b}(1)$$

Where the division property of unity was used to reduce c divided by c to one. Now we can see that if we have a factor repeated in the numerator and denominator we can easily divide out that factor.

Eq 1.3.2 $$\frac{ac}{bc}=\frac{a}{b}$$

This is a very easy idea, but there is one major place that some students struggle with. Consider the example below.

$$\frac{a+c}{b+c}$$

It is tempting to divide out the c's, but there is a critical difference between this statement and the one above. Notice how everything is in terms of multiplication in equation 1.3.2 above. Division is defined within multiplication and so we could easily transfer between the two to isolate our c's and divide them out to one. However, this new statement contains and additive operation. Addition and division are not directly related, in fact they are tiered. No further simplification is possible for this example and we are stuck as is.

Division of Fractions:
Recall equation 1.2.4:

$$\frac{d}{n} = d\left(\frac{1}{n}\right)$$

To start out we will ask the question, what happens if we divide by a number less than one but greater than zero? In math speak:

$$\frac{d}{\left(\frac{1}{n}\right)}$$

Recall what the fundamental question of division is. When we divided we asked how big would one group be if we separated our whole value into n groups. With this same logic we will look at:

$$\frac{d}{\left(\frac{1}{2}\right)}$$

Here we are dividing by half which is (notice) much different than d divided by 2. Again the same process but with new information. How large will one group be if we separated our entire value of d into half of an answer's group? Intuitively if half of a group is d, then the size of a whole group would be twice the size of d, or 2d. Notice that dividing by 2 gives an answer of a half while dividing by a half gives an answer of twice. Let's try the same idea with a third:

$$\frac{d}{\left(\frac{1}{3}\right)} = 3d$$

Because if we placed d into a third of the answers group then the total size of that group would be 3 times the size of d. We can extend this to any number by replacing our fraction with an nth.

Eq 1.3.3

$$\frac{d}{\left(\frac{1}{n}\right)} = d(n) = d\frac{n}{1}$$

The far right side of this argument shows that n can be written with a denominator 1 which is allowed by the first argument of this chapter. We can speculate that dividing by a fraction involves multiplying by what is known as the reciprocal. The reciprocal means to flip the fraction, thus numerator becomes denominator and the denominator turns into the numerator as seen by inspection above. We will utilize this to help explain compound fractions. Compound fractions are fractions divided by fractions that resemble:

$$\frac{\dfrac{m}{w}}{\dfrac{i}{l}}$$

To interpret this we will use the commutative property of multiplication, and equation 1.3.3

$$\frac{\dfrac{m}{w}}{\dfrac{i}{j}} = \frac{m}{w}\frac{j}{i} = \frac{mj}{wi}$$

Thus we have returned to pure multiplication. When dividing compound fractions, flip the second fraction then multiply. Notice how this is NOT cross-multiplication. Cross multiplying is a plague and I will not have it in my book. I cannot rail enough about cross multiplication, don't ever learn it, don't ever use it. Unless, of course, you are using multi-dimensional multiplication.

$$\text{Eq 1.3.4} \qquad \left(\frac{\dfrac{m}{w}}{\left(\dfrac{i}{j}\right)}\right) = \frac{m}{w}\left(\frac{j}{i}\right) = \frac{mj}{wi}$$

Addition of Fractions:

Adding fractions is a generalized case of adding numbers. Before we move on it is important to remember that the definition of addition is not changing at all. We are simply combining two distances on the numberline. Both of these distances just happen to be other than a value you can represent on fingers. However, for us humans to interpret the results (to communicate the answer) we must go through a bit of leg work. Consider the following statement:

$$\frac{a}{b} + \frac{c}{b}$$

Here we are adding amounts of the unit measurement b. Thus if we had 'a' groups of 'b' units and we add 'c' more groups of the 'b' units, how many b units would we have? By simple addition we would have a + c groups of b units.

$$\text{Eq 1.3.5} \qquad \frac{a}{b} + \frac{c}{b} = \frac{a+c}{b}$$

Now many students think that we now have 2 b in the denominator. This is completely silly. Imagine if I give you 5 dollars and then 4 dollars (where 1 dollar is our unit under scrutiny). Would we then have 9 half dollars?

$$\frac{5}{1} + \frac{4}{1} \overset{?}{=} \frac{9}{2}$$

Which means that we have paid a grand total of 4 dollars and 50 cents, less than the initial payment! Did this money grow wings and fly away? Of course not, or else we should wish to buy all our items by law-away and demand all of our paychecks up front and in whole. What a different world that would be. No, the type of unit does not change when we are adding fractions. Thus 5 of a specific type of increment plus 4 of that same type of increment must still be expressed with that same type of increment.

$$\frac{5}{1} + \frac{4}{1} = \frac{9}{1}$$

Thus adding fractions has proven to be easy. This same outcome could be argued another way. Consider using the distributive property in reverse.

$$\frac{a}{b} + \frac{c}{b} = a\left(\frac{1}{b}\right) + c\left(\frac{1}{b}\right) = (a + c)\frac{1}{b} = \frac{a + c}{b}$$

Again we reach the same conclusion. The general rule is when adding amounts of the same unit simply add the numerators. We now move our attention to two fractions with different denominators. Mathematically it would look like this:

$$\frac{a}{b} + \frac{c}{d}$$

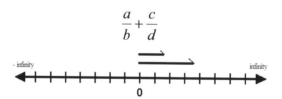

Both of these values are represented by the two vectors on the right. The idea of addition still remains the same. Simply combine the two vectors by placing one tail on the others head. The answer is a value as large as the two amounts combined. But how do we as humans name this new amount so that we understand precisely how much is represented? To solve this problem we need a common measuring stick. Notice in our two amounts (repeated below) that they do not have a similar denominator unit. We have 'a' amount of 'b' increments. We also have 'c' amount of 'd' increments. 'b' and 'd' have nothing in common. Thus we are adding two values of different standards. It's like adding apples and oranges. We can't give one specific number as an answer … unless we redefine each category as fruit!

$$\frac{a}{b} + \frac{c}{d}$$

To get both of these values in the same category we need to have both represented by the same denominator. We can potentially make this denominator anything we want. Generally we like to take the easiest path, which is to simply multiply the two denominators together. Thus our desired common denominator will be 'b' multiplied by 'd' or 'bd'. The process to change the denominator takes two rules of logic that we have developed. The multiplication property of unity, and the divisional property of unity. Which simply means that 1 group of anything is itself (multiplication) and also that any number divided into itself is 1 (other than zero). We will focus on the first term and change the denominator from 'b' to 'bd'

$$\frac{a}{b} = (1)\frac{a}{b} = \left(\frac{d}{d}\right)\frac{a}{b} = \frac{da}{db}$$

Using a similar logic with the second fraction we find:

$$\frac{c}{d} = (1)\frac{c}{d} = \left(\frac{b}{b}\right)\frac{c}{d} = \frac{cb}{db}$$

Notice that if we wanted to we could reduce both of these fractions by dividing out the common factor and we would return to our original notation. In this way we have not changed the actual value of the fraction but only made it a more desirable form. Substituting our new forms into the original statement

$$\text{Eq 1.3.6} \qquad \frac{a}{b} + \frac{c}{d} = \frac{da}{db} + \frac{cb}{db} = \frac{da+cb}{db}$$

Once we found the common denominator we simply then added the numerators, just as before. Subtraction we will remember is defined by addition and thus follows the same argument.

Numerical Examples.

1) $\dfrac{45}{9}$

- We would like to reduce this fraction
- The first thing is to break both numerator and denominator down into their unique prime factorizations.

$$\frac{45}{9} = \frac{(3)(3)(5)}{(3)(3)}$$

- Now we can clearly see two sets of 3 on both numerator and denominator.
- We can divide these sets out like so.

$$\frac{(3)(3)(5)}{(3)(3)} =$$

$$\frac{3}{3}\left(\frac{3}{3}\right)\frac{5}{1} =$$

$$(1)(1)(5) = 5$$

- Notice that we reduced 3 divided by 3 to 1 and not 0. Some students make this mistake
- Notice also that 45 divided by 9 is 5, exactly as we find through this formal process.

2) $\dfrac{2}{3}+\dfrac{2}{5}$

- To make this addition we first need to find a common denominator. The common denominator in this case will be 3(5)=15
- We set about changing our fractions by multiplying by a funny way to say one.

$$\dfrac{2}{3}+\dfrac{3}{5}=$$

$$(1)\dfrac{2}{3}+\dfrac{3}{5}(1)=$$

$$\left(\dfrac{5}{5}\right)\dfrac{2}{3}+\dfrac{3}{5}\left(\dfrac{3}{3}\right)$$

- Carrying out our multiplications we find our common denominator.

$$\left(\dfrac{5}{5}\right)\dfrac{2}{3}+\dfrac{3}{5}\left(\dfrac{3}{3}\right)=\dfrac{10}{15}+\dfrac{9}{15}$$

- Notice how if we wanted to reduce these fractions we would have our original statement, thus we haven't changed the value of either of these fractions, we have only changed the shape.

$$\dfrac{10}{15}+\dfrac{9}{15}=\dfrac{19}{15}$$

3) $\dfrac{\frac{5}{6}}{\frac{1}{4}}$

- This is known as a compound fraction.
- Our first step is to define division by multiplication like so:

$$\dfrac{\frac{5}{6}}{\frac{1}{4}}=\dfrac{5}{6}\left(\dfrac{4}{1}\right)$$

- Now simply multiply across to find our final answer.

$$\dfrac{5}{6}\left(\dfrac{4}{1}\right)=\dfrac{20}{6}=\dfrac{10}{3}$$

- Notice how we did not cross multiply, we multiplied across. In fact never ever utter the word cross multiply again. It is a horrible concept that has no purpose in algebra.
- NEVER CROSS MULTIPLY
- I can't tell you how frustrated I get when good students hold on to this disgusting idea. It is a plague. (I'm usually very tolerant of mathematical interpretations but this 'short-cut' just stinks and causes tremendous problems) I implore you drop it now.

The Calculus Direct

Chapter 1.4 - Simplifying Exponential expressions

Recall that repeated addition of a specific number is termed multiplication. This was prompted by the laziness of mathematicians who developed this shorthand for expedience sake. Working this idea one step further we ask, what would happen if I were to have repeated multiplication of the exact same number? Again mathematicians developed a shorthand system to account for this special case.

We define the exponential as repeated multiplication of the same idea. One can think of a cell dividing as an example, but remember to pull away from concrete examples and focus instead on some malleable idea that could be anything.

$$(x)(x)$$

Here we see the idea of x multiplied by another x. Just as in multiplication we count the number of times the procedure is used. In this case twice, so we define x multiplied by x as x to the second power.

$$\text{Eq 1.4.1} \qquad (x)(x) = x^2$$

Lets define the parts of this definition of the exponential expression. The full script (in this case x) we call the base. It is the variable that will actually be multiplied by itself. The superscript is termed the exponent or the power. It is the amount of times the base is multiplied by itself. Extending this argument by multiplying by another x, and then another x, and another we see a pattern emerge.

$$(x)(x)(x) = x^3 \qquad \text{x to the third power}$$
$$(x)(x)(x)(x) = x^4 \qquad \text{x to the fourth power}$$
$$(x)(x)(x)(x)(x) = x^5 \qquad \text{x to the fifth power}$$
$$(x)(x)(x)......(x)(x) = x^n \qquad \text{x to the nth power}$$

The ellipses in the line above indicate that we repeat the pattern a number of times until we reach a total of n x's being multiplied together. We reserve the superscript directly after a symbol as the exponential. Thus we know at a glance how many times a number is multiplied by itself without having to stop and count.

So what we have done is to count how many times a specific idea is multiplied by itself to determine the exponent on that particular idea. We can extend this argument to generalized exponential expressions of the same base multiplied by each other. Consider

$$\left(x^2\right)(x) = [(x)(x)](x) = x^3$$

Notice how we took the definition of each exponential function and simply added up the number of ideas being multiplied together. Furthering our argument:

$$\left(x^3\right)(x) = [(x)(x)(x)](x) = x^4$$
$$\left(x^4\right)(x) = [(x)(x)(x)(x)](x) = x^5$$

As shown, multiplying our original term by another x we will get a larger final exponential expression that increases the original terms exponent by 1. Formalizing to an exponent of n:

$$\left(x^n\right)(x)=[(x)(x)(x)...(x)(x)](x)=x^{n+1}$$

This argument may be further extended by now concentrating on the second multiplicative term. Increasing the second terms exponent by one produces:

$$\left(x^n\right)\left(x^2\right)=[(x)(x)(x)...(x)(x)][(x)(x)]=x^{n+2}$$
$$\left(x^n\right)\left(x^3\right)=[(x)(x)(x)...(x)(x)][(x)(x)(x)]=x^{n+3}$$

This pattern can clearly be seen to increase the second addend of the exponent. Extending all the way to the 2nd exponent equaling some idea p produces our first logical generalization:

Eq 1.4.2
$$\left(x^n\right)\left(x^p\right)=$$
$$[(x)(x)(x)...(x)(x)][(x)(x)...(x)(x)]=x^{n+p}$$

This is known as the Product rule for exponents. When we multiply the same base with different exponents we simply add the exponents together.

Product raised to a power

Let's multiply two general ideas each raised to a power. Then let's raise that product to a power. This happens very often in technical applications. The mathematics looks like this.

$$\left(a^n b^m\right)^k$$

Since 'a' is not equal to 'b' we cannot apply the power rule. Since mathematics is about starting small and trying to observe patterns let's set k=2 and see what transpires.

$$\left(a^n b^m\right)^2=\left(a^n b^m\right)\left(a^n b^m\right)=\left(a^n a^n\right)\left(b^m b^m\right)=a^{n+n}b^{m+m}=a^{2n}b^{2m}$$

We used the commutative property of multiplication, detailed in section 2, to rearrange the multiplication terms into something we could work with. Notice how the exponents contain repeated groups of addition. Wasn't that the definition of multiplication? With this in mind we were able to express the exponents as a multiplication. So even in superscript, mathematical logic still rings true. Extending our arguments to larger numbers:

$$(a^n b^m)^3 = (a^n b^m)(a^n b^m)(a^n b^m) = a^{n+n+n}b^{m+m+m} = a^{3n}b^{3m}$$

Finally extending to the general case

$$\left(a^n b^m\right)^k=\left(a^n b^m\right)\left(a^n b^m\right)\left(a^n b^m\right)...\left(a^n b^m\right)=$$
$$\left(a^n a^n a^n...a^n\right)\left(b^m b^m b^m...b^m\right)=a^{n+n+n+...+n}b^{m+m+m+...+m}=a^{kn}b^{km}$$

Thus we have generalized the power of a product rule easily. It would do well to stop and take a breather and notice how easy these statements were. All we are doing in mathematics is placing complex concepts into small consistent rules that were already proven. Mathematics then is a very easy subject. The one trick you must learn is to think of the mathematics as an interaction of ideas, not the interaction of numbers. Then we transform from a specific problem to a general problem, but more on this later. Formalizing our statement:

$$\text{Eq 1.4.3} \qquad \left(a^n b^m\right)^k = a^{kn} b^{km}$$

Quotient raised to a power

We move now to the power of a quotient. Consider something that looks as so:

$$\left(\frac{d^n}{e^k}\right)^J$$

By the way, the student should be attempting all derivations on their own before they read on. This gets them used to what true mathematicians have to deal with; the unknown. This argument follows the same idea as the power of a product. I will follow the logic of the derivation but trust the student now has the understanding of what to look for.

$$\left(\frac{d^n}{e^k}\right)^2 = \left(\frac{d^n}{e^k}\right)\left(\frac{d^n}{e^k}\right) = \frac{d^{2n}}{e^{2k}}$$

Moving to the third power

$$\left(\frac{d^n}{e^k}\right)^3 = \left(\frac{d^n}{e^k}\right)\left(\frac{d^n}{e^k}\right)\left(\frac{d^n}{e^k}\right) = \left(\frac{d^n d^n d^n}{e^k e^k e^k}\right) = \frac{d^{3n}}{e^{3k}}$$

Extending to the general case we find:

$$\text{Eq 1.4.4} \qquad \left(\frac{d^N}{e^K}\right)^J = \frac{d^{JN}}{e^{JK}}$$

As you can see this is very easy, mathematics is not very intensive once you learn to think correctly.

Quotient rule of exponents

Now that we have defined what exponential notation is, we can proceed to examine how it works. Consider what would happen if we were to have an exponential expression in the numerator and denominator. Mathematically it would look like this:

$$\frac{x^2}{x}$$

This is easily understood by the idea of multiplication and division. Remember that division is the opposite of multiplication. Thus if we multiply x by itself we get x squared. So then if we divide x squared by an x we return back to x. Mathematically:

$$\frac{x^2}{x} = \frac{(x)(x)}{x} = x\frac{x}{x} = x(1) = x$$

This is brought about by the division property of unity in section 2. We extend this argument to larger and larger values of the numerator exponent.

$$\frac{x^3}{x} = \frac{(x)(x)(x)}{x} = x^2$$

$$\frac{x^4}{x} = \frac{(x)(x)(x)(x)}{x} = x^3$$

$$\frac{x^5}{x} = \frac{(x)(x)(x)(x)(x)}{x} = x^4$$

It becomes apparent that the x in the denominator will effectively divide out one of the x's in the numerator. Thus if we extend our exponent to some general number n in the numerator we will see that once we divide out the denominator we are left with one less than the number the numerator started out with.

$$\frac{x^n}{x} = \frac{(x)(x)(x)...(x)(x)}{x} = x^{n-1}$$

Now what would happen if we were to increase the amount of x's in the denominator? Since we are increasing the amount of x's in the denominator then we will have more x's that can divide out more of the x's in the numerator. Our total number of x's left, represented by our answers exponent, would then be the total number of x's in the numerator minus the total number of x's in the denominator, like so:

$$\frac{x^n}{x^2} = \frac{(x)(x)(x)...(x)(x)}{(x)(x)} = x^{n-2}$$

$$\frac{x^n}{x^3} = \frac{(x)(x)(x)...(x)(x)}{(x)(x)(x)} = x^{n-3}$$

$$\frac{x^n}{x^4} = \frac{(x)(x)(x)...(x)(x)}{(x)(x)(x)(x)} = x^{n-4}$$

Finally, we use the argument that no matter how many x's we have in the denominator we simply subtract that number from the amount of x's in the numerator.

$$\text{Eq 1.4.5} \qquad \frac{x^n}{x^k} = x^{n-k}$$

This is called the quotient rule of exponents. Thus when we encounter a quotient with the same base in the numerator and the denominator, then we use this rule to simplify the expression.

Negative exponents
Everything looks great. But sometimes the quotient rule throws a curve ball. Two main objections come up that are very interesting and in fact play out in many situations. The first goes something like this.

$$x^{-p}$$

The Calculus Direct

Wait a tick... how can x be multiplied by itself a negative amount of times? We can't have negative groups, or so we thought, which prompted our search for what a negative multiplied by a negative was. So how will we interpret this statement? Although bewildering, if we only look to the quotient property it quickly becomes apparent.

$$\frac{x^n}{x^k} = \frac{(x)(x)(x)...(x)(x)}{(x)(x)...(x)(x)} = x^{n-k}$$

So what would happen if k was larger than n? Let's do a simple example and find out.

$$\frac{x^3}{x^5} = \frac{xxx}{xxxxx} = \frac{1}{x^2}$$

Recall that what is happening is that one x from the numerator and one x from the denominator are dividing out in a one to one basis and are removed from the statement. So when the denominator contains larger amounts of x than the numerator then the numerator will be used up first, leaving the denominator with the leftovers.

However let's simply apply our understanding of the quotient rule.

$$\frac{x^3}{x^5} = x^{3-5} = x^{-2}$$

Since we have used two known methods that have given us two true (yet different) answers. Those answers then must both be the truth, and we can equate them. Thus:

$$\frac{1}{x^2} = x^{-2}$$

And so the negative exponent isn't so mysterious after all. It just means that we have a deficit of groups of x in the numerator. Thus to make a negative exponent positive, simply switch which position (numerator or denominator) the variable is located at in the fraction.

Eq 1.4.6
$$x^{-n} = \frac{1}{x^n}$$
$$x^n = \frac{1}{x^{-n}}$$

The beauty of mathematics is that it isn't dependent upon us to make sense. Mathematics simply is and we must accept that it is our understanding, and not it, that is flawed.

Zero exponent
Another statement in question is:

$$x^0$$

Yes, it is a variable x raised to a zero power. Now common logic would say that this is simply x multiplied by itself zero times, which must then be zero. For anything added to itself a total of 0 repeated times is nothing. But, it turns out that x raised to a power of zero is 1.

$$\text{Eq 1.4.7} \qquad x^0 = 1$$

Just as a negative multiplied by a negative gives a surprising answer, so does the zero exponent rule. But we can quickly prove our unintelligent babble by a quick application of the quotient rule of exponents.

$$\frac{x^n}{x^k} = \frac{(x)(x)(x)...(x)(x)}{(x)(x)...(x)(x)} = x^{n-k}$$

We simply state that n is equal to k. and this gives us.

$$\frac{x^n}{x^n} = \frac{(x)(x)...(x)(x)}{(x)(x)...(x)(x)} = x^{n-n} = x^0$$

But we haven't solved our problem, we still don't know what x raised to a power 0 is. But don't we and haven't we? What is anything divided by itself? Well it's 1 of course.

$$\frac{x^n}{x^n} = 1 = x^{n-n} = x^0 = 1$$

There we are! Anything raised to a power 0 is 1. Well again, not quite. There is one glaring deficiency for this rule, when x is equal to 0. Remember we can't divide by zero... yet.

Numerical Examples:

One spot a lot of mistakes crop up is the placement of the subscript. Another is the addition of exponential terms. Study the following examples.

1) $7x^3 = 7(x)(x)(x)$
- This does not mean 7 multiplied three times and x multiplied three times. This only means x multiplied by itself three times.
- The exponent is only touching the x. Thus the x is the only thing affected by the exponent.
- To get 7x multiplied by itself three times we need to have the exponent touch both ideas with the use of parenthesis as below.

2) $(7x)^3 = 7^3 x^3 = 343(x)(x)(x)$
- You may remember seeing this as the power of a product rule.
- Notice the difference between example 1 and example 2
- Remember that exponents are a result of repeated multiplication. Thus addition of terms have no effect on exponents. However repeated addition does have a result of multiplication.
- Thus: Addition breeds Multiplication, Multiplication breeds Exponents. Examples follow:

3) $(x)(x)+(x)(x)=x^2+x^2=2x^2$

- Notice that multiple groups of multiplication caused a shorthand notation of exponents.
- Then multiple groups of addition cause the shorthand notation of multiplication.
- Thus addition groups cause multiplication and multiplication groups cause exponents.
- Remember x plus x is two x. x times x is x squared. Mathematically:

$(x)(x)=x^2$
$(x)+(x)=2x$

4) We would like to simplify the following statement:

$x^2+3x+2x^2-4x+3$

- We have to recall that the variable x stands for a specific number. And that number is repeated each time we see x. However, x and x^2 are not the same number.
- Consider 3 and 3^2. These are not the same number, obviously 3 is three and 3^2 is nine. Thus even though each is defined upon a 3, their value is not the same.
- It is crucial to recall that multiplication is shorthand for repeated groups of addition. Following this standard we will rewrite our statement.

$$\left[(x)(x)\right]+\left[x+x+x\right]+\left[(x)(x)+(x)(x)\right]+\left[-x-x-x-x\right]+3$$

- This is the exact same statement, only with the x's reduced down to their respective multiplication or addition.
- Take some time to verify this fact before moving on.
- Our next step is to see what is common and what is not. Remember that multiplication is only valid if we are adding exact groups. Thus (x)(x) and x are not exact groups and do not validate a shorthand multiplication combination.
- Rewriting our equation again.

$$\left[(x)(x)\right]+\left[(x)(x)+(x)(x)\right]+\left[x+x+x\right]+\left[-x-x-x-x\right]+3$$
$$3\left[(x)(x)\right]+\left[-x\right]+3$$
$$3x^2-x+3$$

5) $\left(\dfrac{\left(a^2b^3\right)^2}{b^2}\right)^{-1}$

- This looks pretty hard but we will see that with small applications of our rules this is very easy.
- First applying the power of a product rule

$$\left(\frac{\left(a^2b^3\right)^2}{b^2}\right)^{-1} = \left(\frac{a^4b^6}{b^2}\right)^{-1}$$

- Now we invoke the quotient rule applied to our b variable.
- Notice that a does not play a role in this because our quotient rule is only applicable to the same base.

$$\left(\frac{a^4b^6}{b^2}\right)^{-1} = \left(a^4b^4\right)^{-1}$$

- This looks much more manageable
- Applying the power of a product rule again.

$$\left(a^4b^4\right)^{-1} = a^{-4}b^{-4} =$$

$$\frac{1}{a^4b^4}$$

- Where the negative exponent rule was used.
- Next we try numbers

6) $\dfrac{357^3}{357^4}\left(357\right)^2$

- We could trudge through this algebraically, or we can use the power of algebra
- Notice that 357 is our base, thus we can use our rules
- First the quotient rule

$$\frac{357^3}{357^4}\left(357\right)^2 = \left(357\right)^{-1}\left(357\right)^2$$

- Next the product rule where we add up our exponents

$$\left(357\right)^{-1}\left(357\right)^2 = \left(357\right)^{-1+2}$$

$$= 357$$

- And we didn't even have to use a calculator.

Chapter 1.5 — Polynomial Terms

Recall that exponentials are repeated groups of multiplication. Recall further that multiplication is repeated groups of addition. With these definitions in mind we look toward further ponderings of exponentials. Consider the following mathematical statement:

$$x^n + x^n$$

Notice how we are adding multiple groups of x raised to the power n. So how would we combine these? One should remember that repeated addition is defined as multiplication.

$$x^n + x^n = 2x^n$$

What we have just developed is known as a term. A term is a collection of variables and numbers through any mathematical operation other than addition. Recall that subtraction is defined within addition, thus subtraction is illegal as well. The terms we will be dealing with are simple polynomic terms. Terms of the type

$$ax^n$$

Where a is defined as the coefficient, x is known as the variable, and n the power or order of the term. The coefficient is a counting number, in the last example it would be 2. We can manipulate terms through the same power of logic that we can manipulate pure numbers. We can add, subtract, multiply, divide, really anything. It is no different than the operation of pure numbers because when we developed these operations we considered malleable ideas. Our payoff for this algebraic approach is easy transition to complex topics with little new leg work.

Multiplying terms

The first combination of terms is multiplication:

$$\left(ax^n\right)\left(bx^p\right)$$

ax^n is simply 'a' multiplied by x raised to the nth power, likewise bx^p is 'b' multiplied by x to the pth power. We can then use the commutative property of multiplication to rearrange our statement.

$$(ax^n)(bx^p) = (a)(x^n)(b)(x^p) = (a)(b)(x^n)(x^p) = abx^{n+p}$$

So, when we need to multiply terms together, we simply multiply the coefficients and add the exponents. Of course we can only add exponents when we have the same base variable. If the base variables are different then we can't combine the exponents because of the definition of exponents. Different based terms would look like this mathematically:

$$(ax^n)(by^p) = abx^n y^p$$

Algebraically this is as far as we can go under our assumptions.

Combining like terms

Many times we will be adding groups of terms with the same powered exponent. Although this may seem new this operation will turn out to be the mathematical argument of multiplication, namely repeated groups of addition. However, rather than discount the subject, I have watched enough students struggle with this concept to merit further clarification. The first example that we studied in this section was an example in combining like terms. Combining like terms is simplifying similar groups of addition into multiplication. We can quickly remember how to combine through multiplication. Consider

$$x + x = 2x$$

We can move further into this by adding sets of multiplication. Consider:

$$2x + 4x = (x + x) + (x + x + x + x) = 6x$$

Where we have split up the terms into their respective definitions then noticed that we were simply adding all of our x's together for a total of 6. This is known as combining like terms. Notice that all of the x's shared the same power and could therefore be considered the same number. But what would happen if we wished to add different powers of x together like so.

$$x^2 + x$$

Can we combine these two terms? No, absolutely not. If the reason is not apparent then a quick revert to our definition of exponents will help us.

$$x^2 + x = (x)(x) + x$$

Quickly we see that we don't have exact groups of addition. Because of this lack of similarity in the terms no simplification is possible. Through this argument it is shown that we can only combine terms of the same power. Cubed terms with cube terms, squared with squared, nth with nth. Numerical examples will help with this.

Numerical Examples.

1) $2x^2 + 3x - 4x + 8x^2 + 3x^3 + x$

- Since we are only allowed to combine terms with the same power this problem is straight forward

$2x^2 + 3x - 4x + 8x^2 + 3x^3 + x =$

$3x^3 + 2x^2 + 8x^2 + 3x - 4x + x =$

$3x^3 + 10x^2$

2) $\left(2x^2\right)\left(x^3 + 4x + 2\right)$

- We will use the distributive property to multiply the term on the left into the summation on the right.

$$\left(2x^2\right)\left(x^3 + 4x + 2\right) = \left(2x^2\right)x^3 + \left(2x^2\right)4x + \left(2x^2\right)2$$

- Now multiplying our coefficients and our variables together.

$\left(2x^2\right)x^3 + \left(2x^2\right)4x + \left(2x^2\right)2 =$

$2x^5 + 8x^3 + 4x^2$

Chapter 2.1 - Equations in one variable

Mathematics exists as statements. These statements maintain relationships with each other and these relationships can be exploited to procure hidden information. One such relationship encountered frequently in mathematics is the equation.

The Equals sign ($=$) is the basis of the equation and states complete convergence of value between the two sides of the symbol. Two entities can only be mathematically equal if they occupy the exact same point on the numberline. When we say two things are equal mathematically we can logically interchange the ideas in separate equations. So rigorous is the algebras that if two ideas cannot be proven absolutely exact then we cannot use $=$ as a relationship symbol and must use $\approx$ which is the "almost" equals sign. The astute student could have a little fun on their exams equipped with this sign and a convincing argument of scale.

Below we designate a relationship that b and c hold towards each other with a mathematical equation. We do not know what either of these variables are, they are simply ideas. However, we do know that they occupy the same exact point on the numberline because our equation says so.

$$b = c$$

We have devoted extensive time to the visualization of addition and multiplication and we will employ those skills presently. Also recall that subtraction is defined by addition and thus any results we find will be equally applicable towards subtraction.

Recall that addition moves the idea a prescribed number of units along the numberline. Thus if we add some number 'n' to b then the position of the left side of the equation moves n units on the numberline. Since the idea b+n no longer coincides with c, the relationship is no longer an equation. We must replace our equals sign with a not-equals sign.

$$b + n \neq c$$

So the question we are confronted with is how to make these two ideas equal again? The main goal is to get the left side of our relationship to occupy the same point as the right side of our relationship so that we may again call both sides equal. Obviously the only sure fire way to do this is to add the same number of units 'n' to c also.

$$b + n = c + n$$

We have brought both sides back into convergence and again can use the equals sign to designate the relationship between them. Thus, if we add anything to one side of an equation we must add that same amount to the other side of the equation to maintain equality. Here our dependence on ideas and variables rather than numbers pays dividends. If we had instead focused on numbers then we would not be able to come to such a nice observation.

Looking at multiplication, and recalling that division is defined within multiplication. Suppose we again start with two ideas that are equal to each other.

$$b = c$$

If we were to multiply the left side by some number 'n' we would lose the identity by the argument above. To maintain the equality we must again repeat the same manipulation to the right side. Generally this involves multiplying the right side by 'n' also.

$$nb = nc$$

Now we keep the validity of the equation and can continue on our way. Thus once given an equation in algebra, we must manipulate both sides at the same time. We find that any time we have an equation and we introduce new information to one side of the equation, we must invariably manipulate the other side of the equation as well.

How to read mathematics

Mathematics is an international language, and certain grammatical issues plague mathematics as much as any language. When we write mathematics we follow a structured hierarchy of performing operations. Our established quota for writing mathematics so that everyone else can follow is GEMA which is an acronym for:

Grouping symbols – Parenthesis, brackets, divisor lines

Exponents

pure **M**ultiplication – Notice that division is defined within multiplication however a divisor line can act as a grouping symbol, hence the need for 'pure' multiplaction examples will follow in the numerical examples

Addition

By following this accepted traditional notation we ensure that others can read our mathematical scribbling long after we've passed on. Numerical examples will help clarify this but study the theory first.

Solving equations

The point of mathematics is to solve for an unknown quantity. Many times we encounter a mathematical equation with an unknown variable. We wish to know at what numerical value of this variable will the relationship become true. To find this unknown variable we must isolate it on one side of the equals sign. Once this is accomplished then the idea on the other side of the equals sign would be the value that our variable must be equal to.

Isolating a variable is not a very hard process. We will simply be performing operations to pull everything else away from our variable. Everything we do however must be repeated on the other side of the equation so that we maintain our equality as shown in the arguments above. Our ultimate goal, in fact the point of algebra, is to find a statement that matches:

$$x = idea$$

If this idea on the right hand side of the equation does not contain an x, then we have solved for x explicitly. If the idea still contains an x then we have only solved the equation implicitly. The mathematician always strives for the explicit answer. To do this we will utilize all of our mathematical ponderings developed so far. Numerical examples follow.

Numerical Examples:

1) $2x + 5 = 7$

- We would like to find the value of x that makes this statement true
- To Do this we will strip away the excess off of x
- The first layer we need to remove is the addend

$$2x + 5 - 5 = 7 - 5$$

$$2x + 0 = 2$$

- We noticed that the opposite operation of addition was subtraction
- Once subtracted we now have separated our constants from variables
- We now move to isolate our variable by the opposite operation of multiplication

$$\frac{2x}{2} = \frac{2}{2}$$

$$x = 1$$

- We are allowed to divide by two on the right hand side if we divide the other side as well
- We have isolated x and found the exact number we need to achieve a true statement. We can check our answer by plugging it in to our original equation

$$2(1) + 5 = 7$$

- Since this statement is true we know that our answer for x is correct

2) $$\frac{2x - 5}{3} = 2x$$

- We have x on both sides, but we wish to have all x's on one side.
- To solve this we first need separate our variables from our constant.
- There are many ways of achieving this. Below is one option

$$3\left(\frac{2x - 5}{3}\right) = (2x)3$$

$$2x - 5 = 6x$$

- We chose to multiply both sides by 3 so as to remove that pesky fraction
- We now can remove our 2x from the left side by subtracting a number that makes our left side 0x.

$$2x - 5 - 2x = 6x - 2x$$

$$-5 = 4x$$

- Notice how the -5 simply falls down because it has not interacted in any way.
- -5 does not contain an x and therefore is not a common term and cannot be affected by adding or subtracting x's.
- Now reducing our 4x down to x is easily accomplished by dividing by 4
- Always remember to do the same operation to both sides.

$$\frac{-5}{4} = x$$

3) $$3(2x + 3) - 2(2x + 3) = 4$$

- again we need to isolate our variable x on one side of the equation.

- Before we jump headlong into this problem and distribute, let's stop and think
- Notice how we have 3(idea) – 2(idea)
- Obviously 3 groups of an idea minus two groups of an idea is just 1 group of an idea.

$$3(2x+3) - 2(2x+3) = 4$$

$$2x + 3 = 4$$

- Notice that we did nothing to the right hand side because we didn't introduce any new operations, only consolidated what was already there.
- Finishing up just like in example 1.

$$2x = 1$$

$$x = \frac{1}{2}$$

4) $2x = 3 - \pi x$

- Again we have x on two different sides, our first task then is to get both on one side
- We choose to put all x terms on the left hand side.
- How then do we make the right hand side have 0x?

$$2x + \pi x = 3 - \pi x + \pi x$$

- Again we have to repeat our new manipulation to both sides.
- Notice that we have both x terms on one side, how then do we achieve a single x?
- By the reverse distributive property

$$x(2 + \pi) = 3$$

- We now wish to isolate x by itself. Simply do the opposite of what is happening to x
- Namely being multiplied (thus opposite is division)

$$x = \frac{3}{2 + \pi}$$

5) $$x = \frac{2(5-3)}{3+6-7} + 3^2 \left(\frac{2-4}{3(2+2-5)} \right)^{-1}$$

- Well… at least x is already isolated
- We will need to use our accepted GEMA approach for this.
- Grouping symbols indicate that they must be approached first, grouping symbols are parenthesis and divisor lines.
- First let's simplify the statements in numerators and denominators.

$$x = \frac{2(2)}{9-7} + 3^2 \left(\frac{-2}{3(4-5)} \right)^{-1}$$

- Notice how in the first term on the right we subtracted 3 from 5 first, this is because we must complete the parenthetic statement before multiplication as prescribed by GEMA.
- Thus we must use GEMA even inside of grouping symbols. GEMA works at several layers.
- We still haven't finished our simplifying of the numerators and denominators so carrying on.

$$x = \frac{4}{2} + 3^2 \left(\frac{-2}{3(-1)} \right)^{-1}$$

$$x = \frac{4}{2} + 3^2 \left(\frac{-2}{-3} \right)^{-1}$$

- Now that the numerator and denominators are simplified we can carry out the division. (note: division is defined as multiplication and should come before addition in GEMA however the divisor line is defined as a grouping symbol, and must be taken as separate parts until there is no more simplification possible.)

$$x = 2 + 3^2 \left(\frac{2}{3} \right)^{-1}$$

- Now that all grouping symbols have been simplified we move to exponents.

$$x = 2 + 9 \left(\frac{3}{2} \right)$$

- Exponents have now been dealt with so we can now move to multiplication.

$$x = 2 + \frac{27}{2}$$

- To add fractions with different denominators we need to find a common denominator.

$$x = \frac{4}{2} + \frac{27}{2} = \frac{31}{2} = 15.5$$

- It is very important to always follow GEMA even when simplifying grouping symbols.

Thus through these examples we have developed a general procedure:
1) Clear fractions by multiplying both sides by a common denominator
2) Simplify both sides
3) Separate our variables and constants onto opposite sides of the equation
4) Isolate our variable

Chapter 2.2 – Functions in one variable

Functional notation is a concise mathematical way to convey information of generalized formulas. We call the function itself the output and the parenthetic variables the input. The input variables are the things we can easily decide upon and control. When we decide on a set of input variables we then want to evaluate the function at those specified inputs. Consider the function defined as:

$$f(x) = 2x + 3$$

Pronounced 'f of x equals two x plus three.' x is our independent variable and f(x) is our actual function. We know that f is a function of x because that is what is inside the parenthesis. If another letter was present on the right side of the equation then f would NOT be dependent upon that. Observe.

$$f(x) = x + c$$

Here again f is a function of x because x is inside of the parenthesis. But f is not a function of c for the simple reason that c is not represented inside of the parenthesis. 'c' is termed a constant coefficient, a funny way of saying a fixed number that doesn't change since it is not an independent nor dependent variable. In essence, c is a fixed number that we just haven't defined yet. But that doesn't mean we can't apply logic to it. More on this later.

Suppose we want to find the value of f for a given x. We do this by simply substituting a chosen value for x every time we see x in our functional equation. For example lets evaluate our function f(x) at x = 4. This would look like:

$$f(x) = 2x + 3$$
$$f(4) = 2(4) + 3 = 11$$

When $x = 4$ then $f(x) = 11$ as shown. Four was an arbitrary choice, it could have been anything. To show this, I've done a few examples with other numbers in rapid succession.

$$f(40) = 2(40) + 3 = 83$$
$$f(3) = 2(3) + 3 = 9$$
$$f(-6) = 2(-6) + 3 = -9$$

Notice how each statement follows the guidelines set by f(x) but substitutes a particular value for each x in the mathematical statement. In this way we can evaluate f(x) for any specific x we choose. To further simplify our answers notation we can employ the ordered pair. The ordered pair lists the substitution value of the independent variable followed by the evaluated answer for the function. It is always listed as (x,f(x)). These three examples would have ordered pairs (40,83) (3,9) and (-6,-9).

Using our definition of f(x) above we can evaluate f for any given input of x. When I say 'any given value of x' this includes other variables as well. Consider evaluating f(x) for x=b:

$$f(x) = 2x + 3$$
$$f(b) = 2b + 3$$

The process hasn't changed, all we do is substitute the value inside of the parenthesis for x throughout the definition. But we can go further.

$$f(b+3) = 2(b+3) + 3 = 2b + 6 + 3 = 2b + 9$$

Nothing has changed! We simply substitute the entire parenthetic statement for our defined variable x. We can go further still.

$$f(2z^2 - 3y + 2) = 2(2z^2 - 3y + 2) + 3 \qquad f(x + \Delta x) = 2(x + \Delta x) + 3$$
$$f(2z^2 - 3y + 2) = 4z^2 - 6y + 7 \qquad f(x + \Delta x) = 2x + 2\Delta x + 3$$

This was a piece of cake. No great genius was needed to complete this. Yet we have just interacted with composite functions. These composite functions are dependent upon other functions who are in turn dependent upon something else. They can be seen as layered upon each other and there is no limit to how many layers one can conceivably have. A composite function is notated like this:

$$f\big(g(x)\big) \qquad \text{or} \qquad (f \circ g)(x)$$

You pronounce this as "f of g of x" on the left or "fog of x" on the right. Both however mean the exact same thing. We will look at an example, but keep in mind that we have already dealt with composite functions so no great effort should be required. Consider the functions:

$$f(x) = ax + c$$
$$g(x) = bx$$

We wish to now give meaning to f(g(x)). We simply introduce g(x) every time we see x in our original function f(x). Thus:

$$f\big(g(x)\big) = a\big(g(x)\big) + c$$

Notice how we have replaced the x with the whole of g(x) while keeping a and c untouched. This is because we are only concerned with replacing x in our function and must leave everything else exactly the same. Inserting our definition of g(x) to finish the example:

$$f\big(g(x)\big) = a\big(g(x)\big) + c = a(bx) + c = abx + c$$

So the evaluation of functions should be no problem. Functions are simply entities like variables, and just like variables we don't know what they are yet. But that doesn't mean we can't use our mathematical logic to manipulate them.

Polynomic Functions of one variable

Consider two different functions. Let's call them $P(\tau)$ and $Q(\tau)$, even though we've changed from f to p and q and from x to τ, the ideas do not change. Do not be confused we are still talking about functions.

$$P(\tau) = a + b\tau + c\tau^2 + \ldots + y\tau^{n-1} + z\tau^n$$

$$Q(\tau) = d + e\tau + f\tau^2 + \ldots + v\tau^{n-1} + w\tau^n$$

Where any of the above constant coefficients may be zero. The two functions above are special cases of functions termed polynomials. A polynomial is a summation of ordered powers of a specific independent variable. Thus anything that fits the idea of $P(\tau)$ or $Q(\tau)$ is termed a polynomial.

Adding and subtracting polynomic functions

Recall from chapter one that subtraction is defined by addition. Thus any argument for addition of polynomials is equally applicable for the subtraction of polynomials as shown by the following mathematical rule:

$$a - b = a + (-b) \qquad\qquad P(\tau) - Q(\tau) = P(\tau) + (-Q(\tau))$$

Notice how we have made the negative sign a part of the second function. This will effectively distribute the negative sign to each term within the second polynomial, which we will argue later. Since we can transform subtraction into addition without any loss of generality we will only cover the addition of polynomials.

Obviously the addition of polynomials is still the simple concept of combining vectors together, just as it was for the addition of pure numbers. Unfortunately with so many variables the pictorial representation tends to become a little complex. Forgoing the pictorial representation we seek a deeper explanation on how to practically add large polynomials. First let's define what the addition of polynomials would look like, consider the following statement.

Eq 2.2.1

$$P(\tau) + Q(\tau) =$$

$$(a + b\tau + c\tau^2 + \ldots + y\tau^{n-1} + z\tau^n) + (d + e\tau + f\tau^2 + \ldots + v\tau^{n-1} + w\tau^n)$$

All that has happened is the definition of each function has been introduced for the respective function $P(\tau)$ and also $Q(\tau)$ with parentheses to distinguish the two. Now, since addition is cumulative, we can remove parentheses and rearrange these terms in any order we see fit. Recalling that we can only add the same powers of τ to each other (see chapter 1.5) our addition can be idealized by rearranging the terms into this particular form.

$$a + d + b\tau + e\tau + c\tau^2 + f\tau^2 + \ldots + y\tau^{n-1} + v\tau^{n-1} + z\tau^n + w\tau^n$$

And finally combining like terms we find:

$$(a + d) + (b + e)\tau + (c + f)\tau^2 + \ldots + (y + v)\tau^{n-1} + (z + w)\tau^n$$

The last line is found by simply pulling out the common factor of τ at each individual power. So it is seen that the addition of functions is simply combining like terms by the power of addition. Notice that this technique can easily be extended to multiple functions added to each other. We simply combine all terms of the same ordered variable.

Another idea to closely scrutinize is what would happen if we were to add a function to itself?

$$P(\tau) + P(\tau)$$

Following our same argument as above, we will only be combining like terms.

$$(a + b\tau + c\tau^2 + ... + z\tau^n) + (a + b\tau + c\tau^2 + ... + z\tau^n) =$$
$$a + a + b\tau + b\tau + c\tau^2 + c\tau^2 + ... + z\tau^n + z\tau^n =$$
$$(a + a) + (b\tau + b\tau) + (c\tau^2 + c\tau^2) + ... + (z\tau^n + z\tau^n) =$$
$$2a + 2b\tau + 2c\tau^2 + ... + 2z\tau^n$$

We should see something that gives us the quintessential warm-tummy feeling. Repeated groups of addition is the definition of multiplication. Every single term inside of our new function is effectively doubled and thus

$$\text{Eq 2.2.2} \qquad P(\tau) + P(\tau) = 2P(\tau)$$

Which is completely in line of our definition of multiplication. Therefore another way we could have completed this problem is by distributing a 2 to each and every term inside the parenthesis. This produces a corollary that should already be apparent but will be mentioned for completeness. Consider multiplying a polynomial by an arbitrary constant.

$$hP(x) = h(a + b\tau + c\tau^2 + \cdots + z\tau^n)$$
$$= ha + hb\tau + hc\tau^2 + \cdots + hz\tau^n$$

This looks exactly like the distributive property developed in chapter one. Multiplying or dividing a polynomial by a constant distributes that constant to each and every term of the polynomial. This is important to remember because we will use this argument during the development of the derivative.

Multiplication of Polynomials
Recall how to multiply groups of addition. (chapter 1.2)

$$(a + b)(c + d) = ac + ad + bc + bd$$

This was found by multiplying every term inside the first set of parentheses by every single term within the second set of parentheses. This has a direct correlation with the function notation as a function is a summation of ordered terms. Mathematically multiplication of polynomials would look like:

$$\text{Eq 2.2.3} \qquad P(\tau)Q(\tau) = (a + b\tau + c\tau^2 + ... + z\tau^n)(d + e\tau + f\tau^2 + ... + w\tau^n)$$

To multiply these polynomials together we must multiply every single term within the first set of parentheses by every single term within the second set of parenthesis. This, as you can see, can become quite tedious. But we must trudge on, some very exciting things are on the way.

$$a(d + e\tau + f\tau^2 + \ldots + w\tau^n) + b\tau(d + e\tau + f\tau^2 + \ldots + w\tau^n)$$
$$+ c\tau^2(d + e\tau + f\tau^2 + \ldots + w\tau^n) + \ldots + z\tau^n(d + e\tau + f\tau^2 + \ldots + w\tau^n)$$

Which, after distributing all of the single terms, and then adding all of our common terms we find our answer.

$$ad + \left(ae + bd\right)\tau + \left(af + be + cd\right)\tau^2 + \ldots + zw\tau^{2n}$$

A very hard task sometimes, but not without importance and a suitable amount of competence is required in this field. The general case of multiplying more than two functions follows the same process and can easily be found. As always numerical examples will appear after the lesson for further clarification.

A very important consideration is what would happen if we were to multiply the same function by itself? Well obviously by our definition of exponents, it would simply be that function squared. Observe:

$$\text{Eq 2.2.4} \qquad P(\tau)P(\tau) = P(\tau)^2$$

Thus we simply square everything inside our function… no that is absolutely wrong. Proceed with the multiplication to see why.

$$(a + b\tau + c\tau^2 + \ldots + z\tau^n)(a + b\tau + c\tau^2 + \ldots + z\tau^n) =$$
$$a(a + b\tau + c\tau^2 + \ldots + z\tau^n) + b\tau(a + b\tau + c\tau^2 + \ldots + z\tau^n) +$$
$$c\tau^2(a + b\tau + c\tau^2 + \ldots + z\tau^n) + \ldots + z\tau^n(a + b\tau + c\tau^2 + \ldots + z\tau^n)$$

And it can quickly be seen that $P(\tau)^2$ will take a bit more elbow grease. But the square of a function is not impossible.

Division of Functions

Here we will just touch on a few needed definitions. We look to give meaning to the two ideas:

$$\frac{P(\tau)}{h} \qquad\qquad \frac{P(\tau)}{Q(\tau)}$$

The first idea on the left corresponds to the division of a polynomial by a constant, something already argued before. To begin, recall that division is defined by multiplication, and finally that multiplication of a function by a constant distributes the constant to each term inside the function. We can carry out this procedure as shown below:

Eq 2.2.5

$$\frac{P(\tau)}{h} = \left(\frac{1}{h}\right)P(\tau) = \left(\frac{1}{h}\right)\left(a + b\tau + c\tau^2 + \ldots + z\tau^n\right)$$

$$= \frac{a}{h} + \frac{b\tau}{h} + \frac{c\tau^2}{h} + \ldots + \frac{y\tau^{n-1}}{h} + \frac{z\tau^n}{h}$$

It was worth our time to explore this distribution of a constant so that we can fully separate the difference between operations involving constants and operations involving polynomials. To divide polynomials we must first replace the polynomial with its definition.

Eq 2.2.6
$$\frac{P(\tau)}{Q(\tau)} = \frac{a + b\tau + c\tau^2 + \ldots + y\tau^{n-1} + z\tau^n}{d + e\tau + f\tau^2 + \ldots + v\tau^{n-1} + w\tau^n}$$

Notice how this operation deals with both division and addition. Remember that we need to find a common FACTOR in both numerator and denominator to reduce those factors to one. If we are able to remove the same factor from numerator and denominator then we could reduce those particular factors to one. Unfortunately under our assumptions we cannot factor the polynomial. Fortunately, however, the ability to factor polynomials is not vital to the calculus. For those purists reading these words you might argue that factoring polynomials is a vital skill to any mathematician. I would agree, but only to a point. As far as a basic understanding of calculus is concerned the factoring of polynomials is an unnecessary skill and as such will not be approached.

Numerical Examples:

1)
$$p(x) = 2x^2 + 3x - 5$$
$$q(x) = 3x + 1$$

- We wish to give meaning to polynomic multiplication and addition.
- First we will tackle multiplication

$$p(x)q(x) = (2x^2 + 3x - 5)(3x + 1)$$

- All we have to do is to multiply every term inside the first set of parenthesis by every term inside the second set of parenthasis
- Just like we did in the multiplication of sums

$$(2x^2 + 3x - 5)(3x + 1) =$$
$$2x^2(3x + 1) + 3x(3x + 1) - 5(3x + 1) =$$
$$6x^3 + 2x^2 + 9x^2 + 3x - 15x - 5$$

- We now just combine like terms (those with the same power)

$$6x^3 + 2x^2 + 9x^2 + 3x - 15x - 5 =$$
$$6x^3 + 11x^2 - 12x - 5$$

- Moving to addition

$$p(x) + q(x) = (2x^2 + 3x - 5) + (3x + 1)$$

- Notice we are adding two ideas that are themselves a summation

- The parenthesis might as well not even be there. We can only combine like terms

$$(2x^2 + 3x - 5) + (3x + 1) = 2x^2 + 6x - 4$$

- Moving to subtraction:

$$p(x) - q(x) = (2x^2 + 3x - 5) - (3x + 1)$$

- Notice how we are subtracting the whole of q(x). Thus we must use our subtraction as defined by addition

$$(2x^2 + 3x - 5) - (3x + 1) = (2x^2 + 3x - 5) + \left(-(3x + 1)\right) =$$
$$(2x^2 + 3x - 5) + (-3x - 1)$$

- After we distribute our negative into our second polynomial we can now approach this as any other addition

$$(2x^2 + 3x - 5) + (-3x - 1) = 2x^2 - 6$$

2) $f(x) = 3x^5 + 2x - 3x^{-1} + 2$

- We will evaluate this function for the following x. x= -2, 5, 3, 0
- Substituting our values into x we find for x = -2

$$f(-2) = 3(-2)^5 + 2(-2) - 3(-2)^{-1} + 2$$

- We now just follow the math, exponents first, then multiplication, finally addition.

$$f(-2) = 3(-32) + 2(-2) - \frac{3}{-2} + 2$$
$$f(-2) = -96 - 4 - \frac{3}{-2} + 2$$
$$f(-2) = -99.5$$

- Heading to the others

$$f(5) = 3(5)^5 + 2(5) - 3(5)^{-1} + 2$$
$$f(5) = 9375 + 10 - \frac{3}{5} + 2$$
$$f(5) = 9386.4$$

- Moving to x=3

$$f(3) = 3(3)^5 + 2(3) - 3(3)^{-1} + 2$$
$$f(3) = 729 + 6 - 1 + 2$$
$$f(3) = 736$$

- These two examples followed the same process and should be no problem
- However when we introduce a value of zero we find a problem

$$f(0) = 3(0)^5 + 2(0) - 3(0)^{-1} + 2$$
$$f(0) = 0 + 0 - \frac{3}{0} + 2$$

- Notice that 3 divided by 0 is an undefined answer. So we can't evaluate this function at x=0.
- We say that x=0 is outside of the function's domain.
- Next chapter we will find a way to scoot past this embarrassing result.

3) $$\frac{x^2 + x}{(x+1)(y^2)}$$

- Notice that we can't simply divide out an x from the top and bottom, because we have addition in the numerator and the denominator. Division and addition are not directly related (recall that multiplication is based upon addition) and cannot be intermingled.
- We must then figure out a way to create multiplication and division
- What would happen for instance, if we factored out the common x on the numerator?

$$\frac{x(x+1)}{(x+1)(y^2)}$$

- Now we have the something. If you don't see it now I will use the commutative property of multiplication and hopefully something will jump out at you.

$$\frac{x(x+1)}{(x+1)(y^2)} = \frac{x}{y^2}\frac{(x+1)}{(x+1)}$$

- Now we can clearly see the opportunity. Recall the divisional property of unity.

$$\frac{x}{y^2}\frac{(x+1)}{(x+1)} = \frac{x}{y^2}(1) = \frac{x}{y^2}$$

- In conclusion remember that we need pure multiplication coupled with pure division to reduce a fraction.

Chapter 2.3 Graphing:

Mathematics has a very real weakness. It can only solve for as many variables as it has coupled unique equations to evaluate. We've already learned how to manipulate functions, now we wish to evaluate and actually graph our results. Notice that the following function actually has two variables.

$$f(x) = mx + b$$

Recall that m and b are both numerical constants because they aren't contained in the parentheses. x is our independent variable, defined because it resides in the parentheses, and f is our second variable, our dependent variable. The point is we can't solve for f or x because we don't have enough information. To circumvent this problem we will solve f(x) for every possible x. A bold statement indeed, as to write all of these down would take... well forever. There is an infinite amount of x possible. We obviously don't want to write an infinite amount of equations and their answers as we have become lazy mathematicians and the thought of a never ending task is not a savory one. To solve this infinite equation paradox we will instead resort to drawing a picture, known as graphing. I've heard that a picture is worth a thousand words. In our case a picture is worth an infinite amount of equations.

The idea behind graphing is to evaluate our function for a suitable amount of x. We ask ourselves to suppose x was some specific value of our choosing. What then would be the corresponding f(x) with that particular x? To solve this riddle we would substitute that specific x into our function and evaluate f(x). Then we repeat the process for another x. After going through a few iterations of this process we are left with a set of ordered pairs. These ordered pairs are then used to plot a picture on a two dimensional Cartesian co-ordinate system.

A Cartesian co-ordinate system is two numberlines set orthogonal to each other, one horizontal and one vertical. The horizontal axes track the independent variable (x) while the vertical axis maps the dependent variable (f(x)). We plot our ordered pair at the orthogonal (perpendicular) intersection of a projected line originating from the ordered pair values on their respective axis. After an acceptable number of points have been plotted we can go ahead and connect our points for the completed graph of our function. As always a numerical example will help us understand.

Once the basic process of graphing is understood, we can just rely on a computerized program to actually graph. We live in a modern world and should use those advantages given to us over our ancestors. Some say we should learn to graph with only a pencil in case we ever find ourselves on a desert island. I've always thought that you should be more worried about food, water, shelter, and not being eaten by predators. Any mathematical ponderings could be reinvented on the desert island if and when you had the time to actually sit and graph a function to begin with. As such, do a search on the internet for a graphing tool and you should have no problem graphing any function. Remember, the graph is a plotting of the answer (f(x)) for every supposal (x). We have almost developed everything we need to delve into the calculus. It has been a long trip but only a few more ideas must be understood.

Superposition

When we graph a polynomial we are really graphing a summation of terms. This creates a special case known as superposition. Superposition states that the answer to the whole is the sum of the parts. Suppose we have three functions, shown below, and that f(x) is a summation of the latter two.

$$f(x) = p(x) + q(x)$$
$$p(x) = mx$$
$$q(x) = b$$

The graphs of all three of these functions have been graphed below.

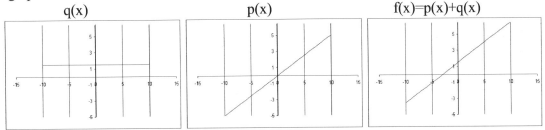

Notice how f(x) has the same shape as p(x) but has been raised along the vertical the exact same amount that q(x) is at a corresponding input value of x. At any particular x, f(x) is the addition of p(x) and q(x) evaluated at that particular x. This shows the power of superposition, that a summation of two functions can be dealt with independently and then added to produce the final result. This is an important idea and one that we will use in the chapters to come.

Numerical Examples

1) $f(x) = 2x - 4$

- To properly graph this function we want to make a few supposals.
- Let's evaluate f(x) for the following x's. x = -5, -2, 0, 1, 10.
- Observe

$$f(-5) = 2(-5) - 4 = -14$$
$$f(-2) = 2(-2) - 4 = -8$$
$$f(0) = 2(0) - 4 = -4$$
$$f(1) = 2(1) - 4 = -2$$
$$f(10) = 2(10) - 4 = 16$$

- Our ordered pairs will be
  $(-5, -14)$
  $(-2, -8)$
  $(0, -4)$
  $(1, -2)$
  $(10, 16)$
- Which is just a way of listing the x value substituted into the function with the corresponding answer to the function.

- With these value of f(x) evaluated at their respective x's we can plot these on the axes of a Cartesian co-ordinate system.
- We will let the horizontal line be our x line and the vertical line be our f(x) line.
- The choice of the horizontal line belonging to our independent variable is arbitrary, it is simply tradition that we chose this configuration.

2) $f(x) = 4x^2 - 2x + 3x^{-2}$

- Once we have defined our function we can then set about evaluating this function for specific x's
- We wish to evaluate for $x = -10, -5, -3, 0, 2, 4$
- All we do is substitute which ever value we want into x and evaluate the function.

$$f(-10) = 4(-10)^2 - 2(-10) + 3(-10)^{-2}$$
$$f(-10) = 400 + 20 + .03$$
$$f(-10) = 420.03$$

- As you can see very easy. Notice how I placed the negative 10 in parenthesis
- This helps remember that the negative sign does play a role.
- Moving through the others in rapid succession

$$f(-5) = 4(-5)^2 - 2(-5) + 3(-5)^{-2}$$
$$f(-5) = 110.12$$
$$f(-2) = 4(-2)^2 - 2(-2) + 3(-2)^{-2}$$
$$f(-2) = 20.75$$
$$f(1) = 4(1)^2 - 2(1) + 3(1)^{-2}$$
$$f(1) = 5$$
$$f(3) = 4(3)^2 - 2(3) + 3(3)^{-2}$$
$$f(3) = 30.33$$

- Using any of the graphing methods to plot.

Chapter 2.4 - Riemann sums:

Let's suppose we have some idea separated into a number of groups, and those groups are comprised of varying amounts of some object. Let's further assume that for some reason we want to know how many objects there are in total. How would we find that out? Well we'd count them up of course.

Let's consider a game we played as children. Spill the contents of a king-size bag of colored candies upon the table and then separate them on basis of color. We have now created partitions (parts of a whole) of candies. So how many were there in the bag in total? Well obviously we don't want to throw all of our candies back together and count them all by our lonesome. We'd want to get everyone to count the number of candies in each partition. Someone counts the red, another counts the green, another the brown, and so on until all of the groups have been counted, even the bits and pieces (depending upon how precise we want to be). Mathematically this might look something like this:

$$g_{red} + g_{blue} + g_{orange} + g_{brown} + g_{black} + g_{yellow} + g_{green} + g_{bits} = g_{total}$$

Where g stands for the number of candies inside each particular group. We can't combine any of these groups through multiplication because none of these groups necessarily contain the exact same number of items. Still we wish that there was an easier way to symbolically write large groups of arbitrary addition. Remember mathematicians are inherently lazy people, anything we can do to write less is considered a good thing.

Removing our thoughts away from candy and instead focusing on malleable ideas we seek to formalize a way to represent large groups of non-repeating addition. What has traditionally been accepted is known as Riemann summation. The use of Riemann, or Sigma, notation allows a mathematician to denote any amount of summation in a concise mathematical statement. Below an example is provided of Riemann notation.

$$\sum_{L=1}^{n} g_L = g_T$$

This states that we will be adding each group of objects (g_L) together to produce the total number of objects (g_T). The number on top of the sigma (n) is the total number of groups for our particular assembly. The statement below sigma, (L=1) designates where we begin our counting of the groups. We can even define how large L should jump for each successive iteration but that is for another book. We will only be concerned with integer summation.

Without further bungling the two statements above mean the exact same thing. Considering red to be our first group, blue to be our second group and so on we find:

$$\text{Eq 2.4.1} \qquad g_1 + g_2 + g_3 + g_4 + g_5 + g_6 + g_7 + g_8 = \sum_{L=1}^{8} g_L = g_{total}$$

The Riemann sum shows its ability when we attach an algebraic expression inside of sigma. One example of this form is shown below.

$$\sum_{L=0}^{n}(L-2)^{L}$$

All we do is replace L with zero (since L starts at zero as defined by the bottom of sigma) at all positions we encounter L in the algebraic expression. Once we evaluate the expression at that value we step the counter up by 1 and set L=1 at all positions inside the expression. Then add those two answers together, then move to two, three, four, and so on until we reach n. A numerical example, as always, is shown at the end of the chapter.

Constant coefficient

Consider what would happen if we were to multiply all of our groups by a single number. Mathematically it would look something like this:

$$\sum_{L}^{n}cx_{L}$$

Which means we are taking the first type of x and multiplying by c. Then we add the second group of x and multiply by c. Then add the third idea of x multiplied by c, and so forth all the way up to n.

$$\sum_{L}^{n}cx_{L}=cx_{1}+cx_{2}+...+cx_{n}$$

Since c is not dependent upon x it will not change as x changes. Looking at our result we see we have a constant number being multiplied by every single term of the sum. Thus by the distributive property in reverse we can easily factor out the common factor:

$$\sum_{L}^{n}cx_{L}=cx_{1}+cx_{2}+...+cx_{n}=c\left(x_{1}+x_{2}+...+x_{n}\right)=c\sum_{L}^{n}x_{L}$$

By this argument, if we have a constant in a Riemann sum, we can place the constant outside of the sigma notation and simply multiply the total by the coefficient rather than each term separately. Formally stating our new idea:

$$\text{Eq 2.4.2} \qquad \sum_{L}^{n}cx_{L}=c\sum_{L}^{n}x_{L}$$

Riemann sum of a sum

Consider taking the sum of two groups. It would look like this:

$$\sum_{L=1}^{n}\left(g(x_{L})+h(x_{L})\right)$$

Carrying out the definition of the Riemann sum:

$$\sum_{L=1}^{n}\left(g(x)_{L}+h(x)_{L}\right)=g(x)_{1}+h(x)_{1}+g(x)_{2}+h(x)_{2}+...+g(x)_{n}+h(x)_{n}$$

Using the commutative property of addition we can mix and match these groups anyway we want. Since I have done this before I will choose to put all the g(x) first and then account for all the h(x). Performing this operation then recombining our sums into respective Riemann sums we find:

$$g(x)_1 + g(x)_2 + ... + g(x)_m + h(x)_1 + h(x)_2 + ... + h(x)_m = \sum_{L=1}^{n} g(x)_L + \sum_{L=1}^{n} h(x)_L$$

Stating this formally:

$$\text{Eq 2.4.3} \qquad \sum_{L=1}^{n} \left(g(x)_L + h(x)_L \right) = \sum_{L=1}^{n} g(x)_L + \sum_{L=1}^{n} h(x)_L$$

Binomial Sum

Before we start this section I want to warn the reader that this is a difficult section. It will go through bewildering arguments, long winded equations, and it will take some elbow grease to get through this section. Take your time, stay as long as needed, and read the whole thing through before giving up. It isn't so hard once you see how everything interacts.

We've learned about the multiplication of the summation of terms, now we wish to approach what it would mean to have one of these summation of terms raised to a power. It would look like this:

$$(a+b)^n$$

To see what this would look like, let's start at n=2 and progress upwards.

$$(a+b)^2 = (a+b)(a+b) = a^2 + 2ab + b^2$$

Recall to square something means only to have a repeated multiplication of itself twice which is what we have done above. Following up with the expansion of the multiplication gives us the right hand side. Moving now to a cubed binomial.

$$(a+b)^3 = (a+b)(a+b)(a+b)$$
$$(a^2 + 2ab + b^2)(a+b) = a^3 + 3a^2b + 3ab^2 + b^3$$

The fourth powered binomial is found in much the same way:

$$(a+b)^4 = (a^3 + 3a^2b + 3ab^2 + b^3)(a+b)$$
$$= a^4 + 4a^3b + 6a^2b^2 + 4ab^3 + b^4$$

Make sure that you are also expanding the polynomial by doing the multiplication and combining of like terms. Also remember that we can only combine addition into multiplication if we have exactly the same variable. $a^2b \neq b^2a$ because when we write all of the a's and all of the b's of the respected expression we get

$$a^2b = aab$$
$$b^2a = bba$$

Which we can see clearly is not the same thing. However $a^2b = ba^2$ because when we write all the a's and all the b's with the associative property of multiplication we find:

$$a^2b = aab$$

$$ba^2 = baa$$

Let's look closely at the pattern of the polynomial that represents the nth power of a binomial. We start with the first term in our binomial and raise it to a power of n. That is the first term in our nth expanded polynomial. The second term of the nth expanded polynomial is the summation of the first term in the n-1 expanded polynomial multiplied by the second term in the binomial with the second term of the n-1 expanded polynomial multiplied by the first term in the binomial. Notice also how we subtract a power from the a and add 1 power of b. Moving to the next term of the nth expanded polynomial we see that we have subtracted another power from a and added another power to b. Also, to get the coefficient we have added the second and third coefficients of the n-1 polynomial due to our multiplication of the binomial.

Rereading the past paragraph I noticed a problem I've had with many of my text books. I've noticed that it only makes sense if you already understand what is going on and is completely unintelligible for all other peoples. To regain my standing with the reader I will explain better.

Let's look directly at the next power of our binomial, or $(a+b)^5$, and call this our nth power. We label the n-1 power then as 5-1=4, or $(a+b)^4$. Since we have already expanded our n-1 polynomial (recall the 4th powered expanded polynomial) we take $a^4 + 4a^3b + 6a^2b^2 + 4ab^3 + b^4$ to be our expanded n-1 (again 4th in this example) powered polynomial.

Remember that any time we wish to multiply groups of addition we need to multiply every single term within the first set of parenthesis with every single term within the second set of parenthesis. Our second set of parenthesis only contains one a and one b. So each term in the first parenthesis can only increase by either a power in a or b.

To find the first term of the nth expanded polynomial (we call it expanded because we no longer simply have a binomial) we take the first term in the binomial (a) and raise it to the nth power (in our case 5)

$$1^{st} \text{ term} = a^5$$

To find the 2nd term we sum the coefficients of the first term in the n-1 polynomial (or a^4) with the second term in n-1 polynomial ($4a^3b$) thus 4+1 =5. We then take away 1 power of a to give (5-1= 4) a^4 and add 1 power to b (0+1=1) b^1. With all the pieces in place we find the second term of the nth expanded polynomial. Another way to look at this is when we multiply the n-1 expanded polynomial with another binomial the only terms that will give us a variable of a^4b is a^4b (The b from the binomial and the a^4 from the first term of 4th expanded polynomial) and $4a^3b(a)$, (which is the a from the binomial and the second term of the 4th expanded polynomial). Thus 1 group plus 4 groups is 5 groups total. All the other multiplications will give us different pairs of powers, thus we don't care about them yet.

$$2^{nd} \text{ term} = 5a^4b$$

Repeating this process but moving to the second and third term of the n-1 polynomial we find the coefficient to be (4+6=10) then subtract 1 more power from a (4-1=3) and add 1 more power to b (1+1=2). Again we only look at the 2nd and 3rd term of the n-1 expanded polynomial because those are the only terms that when multiplied by a or b (the terms in the binomial) will give us our desired pair of powers. Thus the third term is:

$$3^{rd} \text{ term} = 10a^3b^2$$

Repeating this process but moving to the third and fourth term of the n-1 polynomial we find the coefficient to be (6+4=10) then subtract 1 more power from a (3-1=2) and add 1 more power to b (2+1=3). Thus the fourth term is:

$$4^{th} \text{ term} = 10a^2b^3$$

Now that we understand the process I will just list the other terms. Also it would do well to now reread the original explanation and see if you now understand the paragraph.

$$5^{th} \text{ term} = 5ab^4$$
$$6^{th} \text{ term} = b^5$$

To make sure we are multiplying the polynomials together correctly I will carry out the multiplication here.

$$(a+b)^5 = \left(a^4 + 4a^3b + 6a^2b^2 + 4ab^3 + b^4\right)(a+b) =$$
$$a^5 + 4a^4b + 6a^3b^2 + 4a^2b^3 + ab^4 + ba^4 + 4a^3b^2 + 6a^2b^3 + 4ab^4 + b^5$$

It is left to the reader to combine like terms. Once this is accomplished we find the exact set of terms our method above predicted. So as we see, each term in the first parenthesis goes up in power of a and then up in power of b. Combining like terms we see that natural pairs develop and the coefficients are combined.

Now you may wonder why this is in the Riemann sums chapter. The answer is, as we are by now mathematicians, we are inherently lazy people. We wish to find an easy way to list the expansion of the nth power of a binomial without actually multiplying. Well a smarter person than I came up with the following formula. Unfortunately there is a bit of a scandal on who developed this formula. I have heard it was Pascal, I have heard it was Newton, or others unknown. As such I will simply say that without a doubt whoever designed the following mathematical statement is smarter than me.

$$\text{Eq 2.4.4} \qquad (a+b)^n = \sum_{p=0}^{n} \frac{n!}{(n-p)!\,p!} a^{n-p}b^p$$

Where the ! means the factorial developed in chapter 1. In the numerical examples we get to see if this really does describe each term in an expanded polynomial.

Numerical Examples:

1) $\displaystyle\sum_{n=1}^{10} \frac{1}{n}$

- To perform this sum we simply introduce each integer starting with 1 and ending with 10 into the argument.

$$= \frac{1}{1} + \frac{1}{2} + \frac{1}{3} + \frac{1}{4} + \frac{1}{5} + \frac{1}{6} + \frac{1}{7} + \frac{1}{8} + \frac{1}{9} + \frac{1}{10}$$

- Recalling that we have to have common denominators for us to add fractions we know set about achieving common denominators.
- To simplify the process we will only focus on the first two then the third and so on.

$$= \frac{2}{2} + \frac{1}{2} + \frac{1}{3} + \frac{1}{4} + \frac{1}{5} + \frac{1}{6} + \frac{1}{7} + \frac{1}{8} + \frac{1}{9} + \frac{1}{10}$$

$$= \frac{3}{2} + \frac{1}{3} + \frac{1}{4} + \frac{1}{5} + \frac{1}{6} + \frac{1}{7} + \frac{1}{8} + \frac{1}{9} + \frac{1}{10}$$

$$= \frac{9}{6} + \frac{2}{6} + \frac{1}{4} + \frac{1}{5} + \frac{1}{6} + \frac{1}{7} + \frac{1}{8} + \frac{1}{9} + \frac{1}{10}$$

$$= \frac{11}{6} + \frac{1}{4} + \frac{1}{5} + \frac{1}{6} + \frac{1}{7} + \frac{1}{8} + \frac{1}{9} + \frac{1}{10}$$

- Now that we know the process, I convert the denominators and add in one step

$$= \frac{11}{6} + \frac{1}{4} + \frac{1}{5} + \frac{1}{6} + \frac{1}{7} + \frac{1}{8} + \frac{1}{9} + \frac{1}{10}$$

$$= \frac{50}{24} + \frac{1}{5} + \frac{1}{6} + \frac{1}{7} + \frac{1}{8} + \frac{1}{9} + \frac{1}{10}$$

$$= \frac{274}{120} + \frac{1}{6} + \frac{1}{7} + \frac{1}{8} + \frac{1}{9} + \frac{1}{10}$$

$$= \frac{294}{120} + \frac{1}{7} + \frac{1}{8} + \frac{1}{9} + \frac{1}{10}$$

$$= \frac{2178}{840} + \frac{1}{8} + \frac{1}{9} + \frac{1}{10}$$

$$= \frac{2283}{840} + \frac{1}{9} + \frac{1}{10}$$

$$= \frac{7129}{2520} + \frac{1}{10}$$

$$= \frac{7381}{2520} \approx 2.928968$$

2) $$\sum_{0}^{5} (-1)^n \frac{(x)^{2n}}{(2n)!}$$

- This is a special summation
- It designates the Taylor series approximation for the cosine of an angle
- Since x is not dependent upon n we simply treat it as an unknown variable.

$$\left(-1\right)^{0}\frac{\left(x\right)^{2(0)}}{\left(2(0)\right)!}+\left(-1\right)^{1}\frac{\left(x\right)^{2(1)}}{\left(2(1)\right)!}+...+\left(-1\right)^{5}\frac{\left(x\right)^{2(5)}}{\left(2(5)\right)!}=$$

$$1-\frac{x^{2}}{2}+\frac{x^{4}}{24}-\frac{x^{6}}{720}+\frac{x^{8}}{40320}-\frac{x^{10}}{3628800}$$

- I've used the ellipses to show that we simply follow the pattern
- First 0, then 1, 2, 3, 4, and stopping at 5.
- The second line is the truncated series (truncated because we cut the series without expanding to infinity)

3) Find the 3rd power of a binomial using the binomial summation formula.

$$\left(a+b\right)^{n}=\sum_{p=0}^{n}\frac{n!}{\left(n-p\right)!p!}a^{n-p}b^{p}$$

- To use this formula for the 3rd power we let n=3, then compute

$$\left(a+b\right)^{3}=\sum_{p=0}^{3}\frac{3!}{\left(3-p\right)!p!}a^{3-p}b^{p}$$

- Again we simply introduce each evaluation of p and add them up
- First we take p=0, then 1, then 2, and finally 3.
- We know this because p=0 is on the bottom of the Riemann sum.

$$\frac{3!}{\left(3-0\right)!0!}a^{3-0}b^{0}+\frac{3!}{\left(3-1\right)!1!}a^{3-1}b^{1}+\frac{3!}{\left(3-2\right)!2!}a^{3-2}b^{2}+\frac{3!}{\left(3-3\right)!3!}a^{3-3}b^{3}$$

$$\frac{3(2)(1)}{3(2)(1)}a^{3}+\frac{3(2)(1)}{2(1)}a^{2}b^{1}+\frac{3(2)(1)}{2(1)}a^{1}b^{2}+\frac{3(2)(1)!}{3(2)(1)}a^{3-3}b^{3}$$

- Recall that 0! = 1, and 1! = 1
- Simplifying our result we find exactly the same answer as when we did the actual multiplication above.

$$a^{3}+3a^{2}b^{1}+3a^{1}b^{2}+b^{3}$$

Chapter 3.1 - Rate equations

How fast do you drive on the way to the grocery store? How fast can you run a mile? How many times does a wheel spin as a truck drives one mile? All of these questions are rate questions. A rate measures the change of one idea compared to the metered change of another idea. Mathematically we can express the change of the output in regards to the input variable by the rate of change (or rate) formula.

Eq 3.1.1

$$rate = \frac{\Delta f(x)}{\Delta x} = \frac{f(b) - f(a)}{b - a}$$

We use this type of formula because if we want to know how much something has changed then we wish to subtract the value it was at the datum point from its present value. As an illustration of this concept consider your age. How much has your age changed in the past 24 hours if it wasn't your birthday? Your age (in years of course) hasn't changed at all. If it was your birthday in the past 24 hours…Then Happy Birthday from your math book! (But, you've ruined my example. Way to always think about yourself first…) Even though the magnitude of your age at each day is different than zero, the change between the two values is zero. It is this change in value that the rate equation seeks. The triangle used in the rate formula is called delta, a part of the Greek alphabet. Delta has been adopted by the technical community to denote change. The rate of change is equal to the change of the functional value (output) divided by the change in the independent variable (input).

The rate formula allows us to find how much our output has changed given a change in our input. To use the rate equation we will take two arbitrary positions of x and read the value of the related f(x) from the presented graph below. As explained last chapter, when x=a then f(x)=f(a). We don't know what a is nor f(a) in this example, nor do we care. But we will remember that a is a number we choose, then plug into the expression that is f(x) and out pops another number f(a).

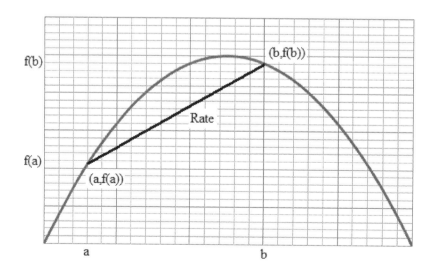

Above our graph has two labeled positions, x= a and x=b, and their corresponding outputs, f(a) and f(b). To find the rate between the two we use the rate formula, the change in output divided by the change in input. What is found by applying the rate formula can be interpreted as the slope of the line connecting the two positions. Above a rate line has been drawn between the two points of interest. If the rate line is steep this can be interpreted as a high rate of change. If the rate line moves up and to the right this can be seen as an increase corresponding with a positive rate. Moving down and to the right is known as a decrease which indicates a negative rate. If the rate line is horizontal then there is no change in the output and the rate is zero.

A better way to write our rate equation is to take into account the fact that we start at some datum point and base the other ordered pair off of this. In this way we would like to find the rate between x and x plus some number. One way to accomplish this is instead of letting the first x=a, we denote the first point of interest as x_0 (a special idea denoting a datum point) and instead of letting the second x=b we will have $x_0 + \Delta x$. What is important to remember is that only our notation has changed, not the idea. The rate is still just the value of a function at a particular input minus the value of the function at another input divided by the change in input. Introducing this new nomenclature we find:

$$\text{Eq 3.1.2} \qquad Rate = \frac{f(x_0 + \Delta x) - f(x_0)}{(x_0 + \Delta x) - (x_0)} = \frac{f(x_0 + \Delta x) - f(x_0)}{\Delta x}$$

We can subtract the common term out of the denominator, but notice that the numerator is functional notation, meaning that we can't simplify the numerator under our current assumptions.

Numerical Examples:

The graph below plots the position (meters) versus the time (minutes) of a person walking down a street and then turning back to walk home.

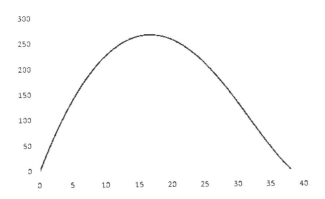

1) What is the rate of change between 0 mins and 5 mins?

- To find the rate of change between the two times we simply apply our rate formula.
- At 0 mins we are 0 meters from the house, at five mins we are 135 meters from the house.
- We will take our datum point to be 0 and our Δx to be (5-0)=5

$$Rate = \frac{f(x_0 + \Delta x) - f(x_0)}{\Delta x} =$$

$$\frac{f(0+5) - f(0)}{5} =$$

$$= \frac{135 - 0}{5} = 27$$

- Our rate of change is shown to be 27 meters per minute.

2) What is the change in distance from the house from 30 minutes to 35 minutes?
- At 30 mins the distance is 135 meters
- At 35 mins the distance is 45 meters
- Using our formula with our datum point to be x=30 and our Δx to be (35-30)=5

$$Rate = \frac{f(30+5) - f(30)}{5}$$

$$= \frac{45 - 135}{5} = -18$$

- Which states that in the five minutes after the thirty minute mark we have reduced our distance from the house by 18 meters/min. (or increased our distance a -18 meters/min, both interpretations work)

3) What is the change in distance from the house from 7 minutes to 25 minutes?

- At 7 mins the distance from the house is 175 meters
- At 25 mins the distance from the house is 210 meters
- Using our formula with our datum point to be x=7 and our Δx to be (25-7)=18

$$Rate = \frac{f(7+18) - f(7)}{18} =$$

$$\frac{f(25) - f(7)}{18} =$$

$$\frac{210 - 175}{18} \approx 1.944$$

- Almost 2 meters per minute. Notice however that we are only taking into account the 25 min mark and the 7 min mark and have missed the hump between these positions entirely.

4) What is the change in distance from the house from 0 minutes to 40 minutes?

- At 0 mins we are 0 meters from the house
- At 40 mins we are 0 meters from the house.
- Using our formula with our datum point to be x=0 and our Δx to be (40-0)=40

$$Rate = \frac{f(0+40)-f(0)}{40} =$$

$$\frac{f(40)-f(0)}{40} =$$

$$\frac{0-0}{40} = 0$$

- Although a true result, this is a dangerous result as the first interpretation would be that we haven't moved at all in the past 40 mins.
- This shows an inherent weakness in the rate formula. Our delta x can't be too large or else we stand to miss pieces of information.
- To keep this from happening we need to use a sufficiently small value of Δx to ensure that no information is lost.
- How small should we take Δx to be to ensure that absolutely no information is lost? We will explore that in the next chapter.

Chapter 3.2 The Derivative

The rate equation, repeated below, allows us to find the change between outputs of a function with regards to the change in the input value.

$$\text{Eq 3.2.1} \qquad rate = \frac{f(x_0 + \Delta x) - f(x_0)}{\Delta x}$$

With this equation we are able to find the change in a phenomenon over time, distance, what have you. As you worked through the example problems last chapter you should have noticed that finding the rate of change between a point and a particular datum point gave a different answer than between a different point and that same datum point. Below a graph helps to highlight the idea that the rate equation based off a common datum point would generally give different answers to the rate equation depending on the choice of the second ordered pair.

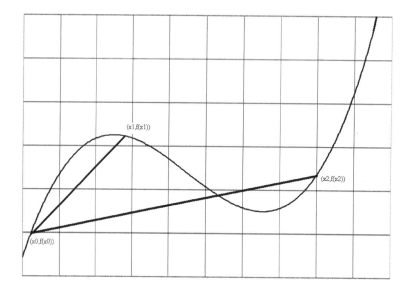

As shown the slope of the two lines are different and therefore, as our interpretation of the slope informs us, the solution to the rate equation is different. To further clarify, think about when you drive. Let's say it takes an hour to drive a total of 20 miles. Using our rate formula we find we traveled

$$rate = \frac{20 miles - 0 miles}{1h} = 20 mph$$

But that doesn't give the whole picture. When you come to a red light you have to completely stop. Your velocity is then 0 miles per hour. When you drive behind the little old lady you only drive 10 mph on the highway. Once you get past her you gun the engine to 90mph out of frustration. So how do we mathematically account for all of these little things that make our trip interesting? How do we find what is happening at a specific moment, which is known as the 'instantaneous change'? If we use our rate equation at a specific instance, meaning no time has passed or no distance covered, then our Δx would be zero. Plugging this into our equation we find:

$$rate = \frac{f(x_0 + 0) - f(x_0)}{0} = \frac{0}{0}$$

The interpretation of the result is that if the input value doesn't change then the output value can't change. Although this is a true mathematical answer this gives no insight to what is happening at the instant, which was precisely the point of using the rate equation to begin with. Since this method failed to give us any useful information, let's try a different approach. Let's do what we can already do, and do it better and better.

Unfortunately we can't define change at a singular point since our definition of the rate equation requires two points. So the question is; which two ordered pairs on the graph will we choose to produce the best estimate of the rate of change that is happening at an exact moment in time? The easy answer is the two points closest together. To illustrate, consider the same graph re-plotted from above with an assortment of x's and their corresponding f(x) value's at different distances from our datum point.

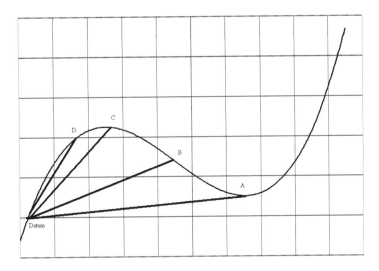

Our first approximation of the rate of change would be to take the difference between the datum point and point a. This is an exaggeration on the first choice of position, but the reader will forgive this exaggeration as it will help clarify the method

$$rate_a = \frac{f(x_0 + \Delta x_a) - f(x_0)}{\Delta x_a}$$

The answer to this difference gives the change of our function over our input for the duration of Δx_a. This is graphed above and clearly our first approximation completely misses the large bump that may be important to us. We need a better approximation that doesn't leave out as much information. To refine our approximation we will then reduce the distance in Δx. This time we will choose x_b. Again following the same approach we fill out the rate formula and get:

$$rate_b = \frac{f(x_0 + \Delta x_b) - f(x_0)}{\Delta x_b}$$

Above the line is drawn (originating at the datum of course) between $(x_0, f(x_0))$ and $(x_b, f(x_b))$. Yet even with this new approximation we have still not accounted for precisely how steep the graph is rising at the datum point and therefore have not reached an acceptable answer to the rate equation. However, our approximation is getting better and this encourages a continuation of the argument. A few more Δx were chosen closer to the datum point and their lines drawn on the graph above. As shown, our approximation gets better at representing all information at and around our specific instance the closer we choose our second point to be with regards to the datum point. To recap what's happening…

We wish to find the 'instantaneous change' or how much our function is changing at a specific point. We found we couldn't take the rate of change at a single point, otherwise we get the nonsensical $\frac{0}{0}$. To solve this problem, we then took better and better approximations of our 'instantaneous' change by choosing smaller and smaller Δx. But we still haven't found the exact answer and we still only get approximations. Approximations are nice but in mathematics it is critical to get exact answers and we now look for a method of achieving that. So now the question is; how small should Δx be to produce the best answer possible?

Without further ado we look now on the method to find the exact 'instantaneous change'. To get the exact answer we must play a bit of mathematical hooky. We can keep doing this approximation thing forever and get better and better approximations that ultimately still leave something to be desired. Or we can model Δx as a variable in its own right that will become zero given the opportunity. I say 'given the opportunity' because setting Δx to zero only causes a problem in the denominator. If we could find a way to eliminate Δx from the denominator then setting it to zero poses no problem. In illustration, consider the following function.

$$f(x) = x^2 + bx + c$$

Taking our rate equation, we find:

$$rate = \frac{f(x_0 + \Delta x) - f(x_0)}{\Delta x} = \frac{\left(x_0 + \Delta x\right)^2 + b\left(x_0 + \Delta x\right) + c - \left[x_0^2 + bx_0 + c\right]}{\Delta x}$$

Recalling that $\left(x_0 + \Delta x\right)^2 = \left(x_0 + \Delta x\right)\left(x_0 + \Delta x\right)$ and carrying out the multiplication, we find:

$$\frac{(x_0 + \Delta x)^2 + b(x_0 + \Delta x) + c - [x_0^2 + bx + c]}{\Delta x} =$$
$$\frac{x_0^2 + 2x_0\Delta x + \Delta x^2 + bx_0 + b\Delta x + c - [x_0^2 + bx + c]}{\Delta x}$$

Notice that we could distribute the negative sign into the right most bracket in the numerator. This allows us to then combine like terms. Performing this operation leaves:

$$\frac{2x_0\Delta x + \Delta x^2 + b\Delta x}{\Delta x}$$

At this point all of the terms in the numerator contain a Δx. We could use the distributive property in reverse to pull out a common factor of Δx like so.

$$\frac{(2x_0 + \Delta x + b)\Delta x}{\Delta x}$$

Notice that if we wished to distribute the Δx back into the parenthesis then we would return to the statement before last. In essence nothing has changed only the way we have written the statement. One very exciting thing this way of writing the equation allows us to do is to use the power of division to eliminate the Δx in both numerator and denominator exactly as argued in equation 1.3.2. What is important to realize is that all logic that worked in chapter 1 still works in our new arguments. Nothing has changed.

$$\frac{(2x_0 + \Delta x + b)\Delta x}{\Delta x} = 2x_0 + \Delta x + b$$

Now remember, mathematically speaking we can't take Δx to be exactly zero. Otherwise we wouldn't be able to divide Δx out of the equation because division by zero is not acceptable in algebra. Instead we will say that Δx is any number we wish and that the answer will get better and better the closer Δx gets to zero, but will never equal zero. This means as Δx approaches zero it becomes so small (infinitesimal) that it does not add any value to our function and can be ignored. If we took the assumption that Δx was infinitesimal then we could neglect any term that contained a Δx. This process is known as taking the limit of a function as Δx approaches zero. But is not zero! In mathematical notation:

$$\lim_{\Delta x \to 0} (2x_0 + \Delta x + b) = 2x_0 + b$$

And we have found the 'instantaneous change' or the 'derivative' at a particular instance. The mathematical argument of the derivative has been completed but what does this mean? How do we use this result to find the instantaneous change?

To find the instantaneous change at a particular moment we need to evaluate this expression at the moment in question. Input your value of x, then maneuver through the solution to the limit equation and you will find the actual number representing the instantaneous change at that particular input value of x. In regards to our statement above, the instantaneous change of our equation is b added to twice our input.

Remember that 'derivative' is just the nomenclature for the process we just developed and the following symbolism defines the derivative.

Eq 3.2.1 $$\frac{df}{dx} = \lim_{\Delta x \to 0} \frac{f(x_0 + \Delta x) - f(x_0)}{\Delta x}$$

We pronounce this as the 'derivative of f(x) with respect to x'. This is the basis of all modern technological thinking. This equation has changed the world.

Notice how we skirt past the embarrassing idea of dividing by zero. We say that we can make Δx smaller and smaller until we are satisfied that it is close enough to zero for us to neglect it. But with the power of the limit we can trust that our answer will be a better than an acceptable approximation, we know that it will be exactly correct! The idea of the derivative is so important that numerical examples will be excluded in this section. Instead take time to dwell on the idea of instantaneous change.

Chapter 3.3 - The study of derivatives

The derivative was developed last chapter and like any other mathematical entity can be manipulated through the use of logic. It is to this purpose that we turn our attention, the algebra of the calculus. Recall the fundamental equation of the derivative.

Eq 3.3.1
$$\frac{df(x)}{dx} = \lim_{\Delta x \to 0} \frac{f(x_0 + \Delta x) - f(x_0)}{\Delta x}$$

Which we pronounce as the derivative of f(x) with respect to x.

The constant coefficient

Suppose we had a function defined by

$$f(x) = (c)(g(x))$$

Where c is a constant coefficient and g(x) is a function in its own right. What would happen if we were to take the derivative of f(x)? By the application of the derivative equation we find:

$$\frac{df(x)}{dx} = \lim_{\Delta x \to 0} \frac{cg(x_0 + \Delta x) - cg(x_0)}{\Delta x} =$$
$$\lim_{\Delta x \to 0} c \left(\frac{g(x_0 + \Delta x) - g(x_0)}{\Delta x} \right)$$

Since c is shared by both terms in the numerator it can be factored by the distributive property of multiplication over addition. C is also not dependent upon x and therefore not dependent upon Δx either. Thus it can be pulled out of the limit all together giving the result

$$c \left[\lim_{\Delta x \to 0} \frac{g(x_0 + \Delta x) - g(x_0)}{\Delta x} \right]$$

Generalizing this result shows that taking the derivative of a function with a constant multiple can be simplified as the constant multiple of the derivative of a function. Formally:

Eq 3.3.2
$$\frac{d(cg(x))}{dx} = c \frac{d(g(x))}{dx}$$

The Power Rule

Assume a function to be any singular term of any power of x. We wish to generalize the derivative. In mathematical notation we set:

$$f(x) = x^n$$

Taking the derivative of this provides:

$$\lim_{\Delta x \to 0} \frac{f(x_0 + \Delta x) - f(x_0)}{\Delta x} = \lim_{\Delta x \to 0} \frac{(x_0 + \Delta x)^n - (x_0)^n}{\Delta x}$$

The generalized multiplication of a binomial has already been established in chapter 2. Application of equation 2.4.4 produces:

The Calculus Direct

$$\lim_{\Delta x \to 0} \frac{(x_0 + \Delta x)^n - (x_0)^n}{\Delta x} = \lim_{\Delta x \to 0} \frac{(x_0^n + nx^{n-1}\Delta x + \frac{n!}{(n-1)!1!}x^{n-2}\Delta x^2 ... + \Delta x^n) - (x_0)^n}{\Delta x}$$

After we subtract the common x_0^n one should notice that every single term has a Δx .

$$\lim_{\Delta x \to 0} \frac{nx^{n-1}\Delta x + \frac{n!}{(n-2)!\,2!}x^{n-2}\Delta x^2 + \cdots + \Delta x^n}{\Delta x}$$

Thus we can factor out a Δx from each of the terms in the numerator and divide out the common Δx from numerator and denominator.

$$\lim_{\Delta x \to 0} \frac{\left(nx^{n-1} + \frac{n!}{(n-2)!\,2!}x^{n-2}\Delta x^1 + \cdots + \Delta x^{n-1}\right)\Delta x}{\Delta x}$$

This eliminates the Δx in the denominator and leaves us with a simplified numerator

$$\lim_{\Delta x \to 0} \left(nx^{n-1} + \frac{n!}{(n-2)!\,2!}x^{n-2}\Delta x^1 + \cdots + \Delta x^{n-1}\right)$$

At this point we can use the trick that Δx is 'close enough to zero'. Taking this assumption forces every term inside the parenthesis to reduce to zero except for the first term. The reason for this is because anything multiplied by an infinitesimal is itself infinitesimal and can be neglected by our argument. In essence we are multiplying by a number just shy of zero which gives a product just an iota larger than zero, which we can neglect by our assumptions. Formalizing the power rule for derivatives:

$$\text{Eq 3.3.3} \qquad \frac{d(x^n)}{dx} = \lim_{\Delta x \to 0} \frac{(x_0 + \Delta x)^n - (x_0)^n}{\Delta x} = nx^{n-1}$$

Of course our rule is derived from an argument that only takes into account any integer greater than or equal to zero. Unfortunately, any term raised to a negative integer power will have to wait until we have a few more tools to work with.

The law of superposition
Recall that a polynomial can be classified as a summation of powered terms of a specified variable.

$$f(x) = a + bx + cx^2 + \cdots + zx^n$$

These ordered terms could be considered a function in their own right. To do this let's set r(x) equal to a, s(x) equal to bx, t(x) equal to cx^2, and so on until we reach the final term. This produces a polynomial of functions rather than a polynomial of variables.

$$f(x) = r(x) + s(x) + t(x) + \cdots + z(x)$$

This assumption allows for two important results. It generalizes the polynomial and also extends the superposition principle beyond single terms and into functions. Before we apply the definition of the derivative to the statement above it would be helpful to make an observation on the derivative itself. One should notice that by equation 1.3.5 we could rewrite the definition of the derivative as so.

$$\frac{df(x)}{dx} = \frac{f(x_0 + \Delta x) - f(x_0)}{\Delta x} = \frac{f(x_0 + \Delta x)}{\Delta x} - \frac{f(x_0)}{\Delta x}$$

The right most side of the equation above is found by separating the statement into two fractions. At this point it is easy to substitute the summation of functions for f(x).

$$\lim_{\Delta x \to 0} \frac{\left(r\left(x_0 + \Delta x\right) + s\left(x_0 + \Delta x\right) + ... + z\left(x_0 + \Delta x\right) \right)}{\Delta x} - \frac{\left(r\left(x_0\right) + s\left(x_0\right) + ... + z\left(x_0\right) \right)}{\Delta x}$$

Since addition has the commutative property, we can mix and match the numerator in any way we wish. So combining the respective terms together we get.

$$\lim_{\Delta x \to 0} \frac{\left(r\left(x_0 + \Delta x\right) - r\left(x_0\right) \right)}{\Delta x} + \frac{\left(s\left(x_0 + \Delta x\right) - s\left(x_0\right) \right)}{\Delta x} + ... + \frac{\left(z\left(x_0 + \Delta x\right) - z\left(x_0\right) \right)}{\Delta x}$$

This shows that the derivative of a polynomial is equal to the summation of the derivatives of each polynomic term taken separately. Generalizing the statement mathematically

$$\text{If } f(x) = r(x) + s(x) + ... + z(x)$$

Eq 3.3.4

$$\frac{df(x)}{dx} = \frac{dr(x)}{dx} + \frac{ds(x)}{dx} + ... + \frac{dz(x)}{dx}$$

Higher Order Derivatives

We now have a basic understanding of how to take the derivative. What would happen, however, if we turned right around and took another derivative of the function? It would do well for us to remember that the derivative of a function is simply another function. It is connected to the original function to be sure, but is still a function in its own right.

It turns out this is perfectly legal. We call this a higher order derivative. We will recall that constant coefficients can be pulled out of the derivative and also that the polynomial may be separated using the superposition principle. Thus we will restrict our attention to a single term of a power of x. We have already gone through the process of taking the derivative of a powered term. So we will only extend on our results and expect the reader to remember how to use the power rule.

$$f(x) = x^n$$
$$\frac{df(x)}{dx} = nx^{n-1}$$

We now look to take the second derivative

$$\frac{d}{dx}\left(\frac{df(x)}{dx}\right) = \frac{d}{dx}(nx^{n-1}) = n\frac{d(x^{n-1})}{dx}$$

Which can be interpreted as the derivative of x to a power of n-1. Notice that the constant term n has been factored out. Using the power rule again at this point produces:

$$n(n-1)x^{n-2}$$

Thus it is perfectly acceptable to take a second derivative from an original function. This is not illegal and in fact is used so many times in technical application that there is a way to denote a second order derivative formally, that method being shown below:

$$\frac{d}{dx}\left(\frac{df(x)}{dx}\right) = \frac{d^2 f(x)}{dx^2}$$

But we don't have to stop there, we could conceivably take as many derivatives as we wish one right after the other. Mathematically we would denote this as:

Eq 3.3.5

$$\frac{d^n f(x)}{dx^n}$$

Where the n signifies that we have taken the derivative n times. This notation is encountered in many mathematical formulas. The Taylor series, for instance, uses an infinite sum of infinite integer derivatives. One might even wonder what would happen if n were not an integer but a decimal or a fraction. Again this is beyond the scope of this book, but the interested student can find a wide range of texts covering the fractional calculus. What is important to know is that the material is not impossible to learn, just like the calculus we learn today. It only takes effort.

Further explorations

We will finish this chapter with a mental exercise. What is the derivative of a variable with respect to itself? In other words what is the derivative of x with respect to x?

$$\frac{dx}{dx}$$

By our intuition anything divided by itself (no matter how small) must be equal to 1. So how would we go about proving this? Let's make a function equal to x and see what happens when the definition of the derivative is utilized.

$$f(x) = x$$

$$\frac{df(x)}{dx} = \lim_{\Delta x \to 0} \frac{x_0 + \Delta x - x_0}{\Delta x} = 1$$

Easy enough. But is our new rule constricted only to x? could we say give meaning to

$$\frac{dg(x)}{dg(x)}$$

Of course we can. We have spent a whole book thinking in ideas. And this idea is simply asking what is the rate that g(x) changes with respect to g(x). We can also look at it as g(x) is both the input and the output. So what is the change in output for a change in input if both output and input are exactly the same? Obviously it would be a 1 to 1 ratio. Thus:

$$\text{Eq 3.3.6} \quad \frac{dg(x)}{dg(x)} = \lim_{\Delta x \to 0} \frac{g(x)_0 + \Delta g(x) - g(x)_0}{\Delta g(x)} = 1$$

As the math shows. Notice that we don't take $g(x_0)$ but rather $g(x)_0$. It might seem confusing but all we are saying is that we take some value g(x) as our datum value $g(x)_0$ which varies from taking some x as our datum point.

It has been said that derivatives are a quotient of infinitesimals. These infinitesimals combine in ways to produce a new and exciting understanding of the world. Many, many things can be done with these infinitesimals, but this book only serves as an introduction. I highly recommend you consult other books and further understand this amazing calculus. Next we will look at a few advanced topics in the calculus.

Numerical Examples

1) Find the derivative of $x^3 + 3x^2 + 2x + 7$ and then find the instantaneous change at x=1

- First we use the law of superposition to break apart our derivative.
$$\frac{d(x^3 + 3x^2 + 2x + 7)}{dx} = \frac{d(x^3)}{dx} + \frac{d(3x^2)}{dx} + \frac{d(2x)}{dx} + \frac{d(7)}{dx}$$
- Notice that the last term (7) could be considered as having a variable x to the 0th power. Introducing this we will have $7 = 7(1) = 7x^0$
- Now we will pull the constant coefficient out of each derivative
$$\frac{d(x^3)}{dx} + 3\frac{d(x^2)}{dx} + 2\frac{d(x)}{dx} + 7\frac{d(x^0)}{dx}$$
- We now apply the power rule for each of the terms
$$\left[3x^2\right] + 3\left[2x\right] + 2\left[1x^0\right] + 7\left[0x^{-1}\right]$$
- Multiplying what we need to and simplifying x to the zeroith we find:
- Notice that the last term will be zero because 0 multiplied by anything finite is 0.

$$3x^2 + 6x + 2$$
- Answering the second part of the problem we insert a "1" into each x to find our instantaneous change

$$\frac{df(1)}{dx} = 3(1)^2 + 6(1) + 2 = 11$$
- As can be seen our function is increasing at 11 units of output for every 1 unit of input at the exact moment that x=1.

2] Take the derivative of $15x^2 + 3x - 8$ and find the instantaneous change at x=2

- Approaching this problem the same way we split up our derivative by the law of superposition.

$$\frac{d\left(15x^2 + 3x - 8\right)}{dx} = \frac{d\left(15x^2\right)}{dx} + \frac{d\left(3x\right)}{dx} + \frac{d(-8)}{dx}$$

- Pulling the constant coefficients out and realizing the third term has a variable x to the zeroith power.

$$\frac{d\left(15x^2\right)}{dx} + \frac{d\left(3x\right)}{dx} + \frac{d(-8)}{dx} = 15\frac{d\left(x^2\right)}{dx} + 3\frac{d\left(x\right)}{dx} - 8\frac{d(x^0)}{dx}$$

- Notice how the negative sign accompanied the 8 in the third term.
- Using the power rule
- The last term may seem strange but we simply apply the power rule without problem

$$15[2x] + 3[1x^0] - 8[0x^{-1}]$$

- Simplifying our result.

$$30x + 3$$

- Evaluating our derivative at x=2 gives our instantaneous change at that particular moment.

$$\frac{df(2)}{dx} = 30(2) + 3 = 63$$

- Thus our function is changing by 63 units of output per 1 unit of input at the exact moment that x=2

Chapter 3.4 Advanced Derivatives

The first attempt at the advanced derivatives will be the powerful chain-rule. Recall that we can have a composite function described in chapter two as:

$$f(g(x))$$

Remember that f(g(x)) is composed of g(x) which itself is composed of x. We want to take the derivative of f(g(x)) with respect to x. Mathematically it looks something like this:

$$\frac{df(g(x))}{dx}$$

But we haven't developed any way of taking a derivative with respect to another variable. All we know how to do is to take a derivative with respect to an independent variable. Recall that g(x) is generally completely different than plain x. We circumvent this problem by remembering how fractions operate. Recall that anything multiplied by one is itself. Further, recall that the derivative of a variable with respect to itself is equal to one. (We developed this in the last section). With these assumptions we can use the following logic.

$$\frac{df(g(x))}{dx} = \frac{df(g(x))}{dx}(1) =$$

$$\frac{df(g(x))}{dx}\frac{dg(x)}{dg(x)} =$$

$$\frac{df(g(x))}{dg(x)}\frac{dg(x)}{dx}$$

Notice how we are now taking the derivative first with g(x), which can be interpreted as the independent variable of f(x). Then we multiply by the derivative of g(x) with respect to its independent variable x.

Eq 3.4.1 $$\frac{df(g(x))}{dx} = \frac{df(g(x))}{dg(x)}\left(\frac{dg(x)}{dx}\right)$$

This is known as the chain rule. It is a very powerful tool that can be applied to many unsavory functions. Now a very exciting thing is possible. Consider if g(x) was itself a composite function composed of yet another function.

$$f(g(h(x)))$$

How would we then find the derivative of f(x) with respect to x? through the same process of course:

$$\frac{df\left(g(h(x))\right)}{dx}(1)(1) =$$

$$\frac{df\left(g(h(x))\right)}{dx}\frac{dg(h(x))}{dg(h(x)}\frac{dh(x)}{dh(x)} =$$

$$\frac{df\left(g(h(x))\right)}{dg(h(x))}\frac{d\left(g(h(x))\right)}{dh(x)}\frac{dh(x)}{dx}$$

As you can see we can just chain as many composite functions along as we wish, hence the chain rule. Mathematically the generalized chain rule would look like:

$$\text{Eq 3.4.2}$$

$$\frac{df\left(g\left(...(z(x))\right)\right)}{dx} =$$

$$\frac{df\left(g\left(...(z(x))\right)\right)}{dg\left(...(z(x))\right)}\frac{dg\left(...(z(x))\right)}{d\,...(z(x))}\,...\,\frac{dz(x)}{dx}$$

The understanding of this should not be a problem even though the equation is complex. As always a numerical example follows after the chapter to solidify the understanding.

Derivative of a product of functions

Consider the function f(x) which is defined as two functions multiplied together.

$$f(x) = g(x)h(x)$$

We wish to take the derivative of this function with respect to x:

$$\frac{df(x)}{dx} = \frac{d\left(g(x)h(x)\right)}{dx} =$$

$$\lim_{\Delta x \to 0}\frac{g\left(x_0 + \Delta x\right)h\left(x_0 + \Delta x\right) - g\left(x_0\right)h\left(x_0\right)}{\Delta x}$$

At this point we are stuck since multiplication and addition are not directly related. In fact one will recall that multiplication is derived from addition. To solve this problem we will have to remember the trick we used when we proved that a negative multiplied by a negative is a positive. We added 0.

$$\frac{g(x_0)h(x_0 + \Delta x)}{\Delta x} - \frac{g(x_0)h(x_0 + \Delta x)}{\Delta x} = 0$$

Of course, anything subtracted from itself is zero. As such if we add the right hand side of the equation into the limit statement we haven't really added anything since it is equal to zero. Using this concept and the commutative property of addition with the original statement produces an intuitively obvious result.

$$\lim_{\Delta x \to 0} \frac{g(x_0 + \Delta x)h(x_0 + \Delta x) - g(x_0)h(x_0 + \Delta x)}{\Delta x} + \frac{-g(x_0)h(x_0) + g(x_0)h(x_0 + \Delta x)}{\Delta x}$$

Now we can clearly see the beginnings of the classical derivative definition. We move forward by factoring out a common factor from each fraction respectivly.

$$\lim_{\Delta x \to 0} \frac{g(x_0 + \Delta x) - g(x_0)}{\Delta x} h(x_0 + \Delta x) + \lim_{\Delta x \to 0} \frac{h(x_0 + \Delta x) - h(x_0)}{\Delta x} g(x_0)$$

Notice that the two terms in the second limit have been switched using equation 1.1.4. The factor from the first limit can be simplified by setting $h(x_0 + \Delta x)$ equal to $h(x_0)$ as Δx goes to zero. Directly adjacent to it is the well known derivative of g(x) with respect to x. The second Limit is also easily distinguished as the derivative of h(x) with respect to x multiplied by $g(x_0)$. Thus we find that the derivative of a product is:

Eq 3.4.3
$$\frac{d[g(x)h(x)]}{dx} = \frac{dg(x)}{dx} h(x_0) + g(x_0) \frac{dh(x)}{dx}$$

Just as it may seem counterintuitive that our solution looks like this, we must remember that we base our logic off the mathematics and not the other way around.

The derivative of a quotient
The derivative of a product was surprising; we shall see that the derivative of a quotient is even more surprising. We wish to take the derivative of a complex function of the type:

$$f(x) = \frac{g(x)}{h(x)}$$

Carrying out the definition of the derivative:

$$\frac{df(x)}{dx} = \frac{d\left(\frac{g(x)}{h(x)}\right)}{dx} = \lim_{\Delta x \to 0} \frac{\frac{g(x_0 + \Delta x)}{h(x_0 + \Delta x)} - \frac{g(x_0)}{h(x_0)}}{\Delta x}$$

We need to subtract the two fractions in the numerator to proceed. The particulars on how to do this was explained by equation 1.3.6. We set about finding the common denominator by multiplying the two denominators together. The common denominator by this method is $h(x_0 + \Delta x)h(x_0)$

$$\lim_{\Delta x \to 0} \frac{\frac{g(x_0 + \Delta x)}{h(x_0 + \Delta x)} \frac{h(x_0)}{h(x_0)} - \frac{g(x_0)}{h(x_0)} \frac{h(x_0 + \Delta x)}{h(x_0 + \Delta x)}}{\Delta x} =$$

$$\lim_{\Delta x \to 0} \frac{\frac{g(x_0 + \Delta x)h(x_0) - g(x_0)h(x_0 + \Delta x)}{h(x_0 + \Delta x)h(x_0)}}{\Delta x}$$

Notice we now have a compound fraction. Using equation 1.3.4 to rewrite the statement gives.

$$\lim_{\Delta x \to 0} \frac{g\left(x_0 + \Delta x\right)h\left(x_0\right) - g\left(x_0\right)h\left(x_0 + \Delta x\right)}{h\left(x_0 + \Delta x\right)h\left(x_0\right)}\left(\frac{1}{\Delta x}\right)$$

Now, we use a little algebraic magic to switch the denominators into a more agreeable form:

$$\lim_{\Delta x \to 0} \frac{g\left(x_0 + \Delta x\right)h\left(x_0\right) - g\left(x_0\right)h\left(x_0 + \Delta x\right)}{\Delta x}\left(\frac{1}{h\left(x_0 + \Delta x\right)h\left(x_0\right)}\right)$$

To simplify the right side of the limit, we notice that as $\Delta x \to 0$ then $h(x_0 + \Delta x) \to h(x_0)$. How we pronounce this is 'as delta x approaches zero, h of x not plus delta x approaches h of x not'. Obviously x not plus zero will be x not, so we are okay there. Notice how this is not the same thing as placing a zero in the denominator, rather we are evaluating h(x) while adding nothing to x. Focusing on the left side of the limit we must use the same trick we used for the derivative of a product. We will add a special type of zero that makes our life easier.

$$\frac{g(x_0)h(x_0 + \Delta x)}{\Delta x h(x_0)^2} - \frac{g(x_0)h(x_0 + \Delta x)}{\Delta x h(x_0)^2} = 0$$

Who said we couldn't get something from nothing? Rewriting our equation by intermingling the terms by the commutative property of addition just as before we find:

$$\lim_{\Delta x \to 0} \frac{g(x_0 + \Delta x)h(x_0) - g(x_0)h(x_0)}{\Delta x h(x_0)^2} - \frac{g(x_0)h(x_0 + \Delta x) - g(x_0)h(x_0)}{\Delta x h(x_0)^2}$$

Notice how our first limit has the classic derivative in regards to g(x) and the second term is the classic derivative of h(x). For those that are having trouble making the second fraction look the same as above, try distributing the negative sign into the numerator of the second fraction and then indentify each term that went into the argument. At this point we can factor out the common factor from each fraction respectively giving:

$$\lim_{\Delta x \to 0} \frac{1}{h(x_0)^2}\left[\frac{g(x_0 + \Delta x) - g(x_0)}{\Delta x}h(x_0) - \frac{h(x_0 + \Delta x) - h(x_0)}{\Delta x}g(x_0)\right]$$

Formalizing by notating with the definition of the derivative:

Eq 3.4.4

$$\frac{d\left(\frac{g(x)}{h(x)}\right)}{dx} = \frac{\frac{dg(x)}{dx}h(x) - \frac{dh(x)}{dx}g(x)}{h(x)^2}$$

And we have successively (after much effort, but not difficult effort) found the derivative of a quotient.

Extending the power rule

In the last section we derived the power rule for any integer greater than zero by using arguments from the binomial theorem. Now we move to make the power rule, well… more powerful, by incorporating all of the negative integer exponents as well. Our main goal is to give meaning to the derivative of a function of the following form.

$$f(x) = x^{-n}$$

$$\frac{df(x)}{dx} = \frac{d\left(x^{-n}\right)}{dx}$$

By now we should know that mathematics is about making something you don't know into something you do know. For instance we know that a negative exponent can be turned positive by moving the term into the denominator as described in equation1.4.6.

$$\frac{df(x)}{dx} = \frac{d\left(x^{-n}\right)}{dx} = \frac{d}{dx}\left(\frac{1}{x^n}\right)$$

Now we can easily apply the quotient rule without any problems.

$$\frac{d}{dx}\left(\frac{1}{x^n}\right) = \frac{\dfrac{d(1)}{dx}x^n - \dfrac{d\left(x^n\right)}{dx}1}{\left(x^n\right)^2}$$

The first derivative in the numerator is quickly seen to be zero (recall the power rule for derivatives with n=0) and the second derivative is quickly calculated using our power rule with exponent n.

$$\frac{0 - \dfrac{d\left(x^n\right)}{dx}1}{\left(x^n\right)^2} = -\frac{nx^{n-1}}{x^{2n}}$$

This next step was formulated in chapter 1. If we have pure multiplication and a common base in both numerator and denominator then we can easily combine them into one base by subtracting the two exponents, notice how the negative sign is not affected by this manipulation.

$$-\frac{nx^{n-1}}{x^{2n}} = -nx^{n-1-2n} = -nx^{-n-1}$$

What a great result! As we can see the result does not vary from the power rule one iota. We have a continuity across the exponents in integers. Formalizing

$$\frac{d\left(x^{-n}\right)}{dx} = -nx^{-n-1}$$

Or we could just simply say that the power rule works for all integer n. In fact, it turns out the power rule works for all real exponents regardless of fractions, decimals, or irrational numbers. The proof of this is not difficult, but since our goal is to only have a basic understanding of the calculus and I have placed a 100 page limit on this book we will not delve into this matter. But the interested student can find the information. Good luck!

Numerical examples.

    1) Take the derivative of $\left(x^2 + 3\right)^4$

- This looks complex and it is, but a quick definition allows for an easy approach

Let
$$f(x) = (x)^4 \qquad f(g(x)) = (g(x))^4$$
$$g(x) = (x^2 + 3) \qquad f(g(x)) = (x^2 + 3)^4$$

- Now we can use the chain rule to easily take the derivative.
- The first portion of the chain rule is seen to be the power rule in disguise

$$\frac{df(g(x))}{dg(x)} = \frac{d(g(x))^4}{d(g(x))} = \frac{d(x^2+3)^4}{d(x^2+3)} = 4(x^2+3)^3$$

- Notice how taking the derivative of $g(x)^4$ with respect of $g(x)$ produces the same effect as taking the derivative of x^4 with respect to x
- Thus our adherence to thinking in ideas has paid off.
- Moving to the second part of the chain rule

$$\frac{dg(x)}{dx} = \frac{d(x^2+3)}{dx} = 2x$$

- The law of superposition was used from last chapter
- Combining both into the total we find:

$$\frac{df(g(x))}{dg(x)} \frac{dg(x)}{dx} = \frac{d(x^2+3)^4}{d(x^2+3)} \frac{d(x^2+3)}{dx} = \left[4(x^2+3)^3\right][2x] = 8x(x^2+3)^3$$

2) Find the derivative of $x^4(x^2-1)$

- Notice that we could easily distribute then take the derivative to find $6x^5 - 4x^3$
- However we wish to test the product rule.

$$g(x) = x^4$$
$$h(x) = (x^2 - 1)$$

- Now we apply our product rule

$$\frac{dg(x)}{dx} h(x) + g(x) \frac{dh(x)}{dx} = 4x^3(x^2-1) + x^4(2x)$$

- Simplifying our result we find

$$4x^3(x^2-1) + x^4(2x) = 4x^5 - 4x^3 + 2x^5 = 6x^5 - 4x^3$$

- Exactly our result above
- Although this may seem arbitrary, a highly powerful tool will soon be based off this approach.

3) Find the derivative of $x^{\frac{5}{2}}$

- Even though we are dealing with radical powers of x, we can still use our powerful power rule.

$$\frac{dx^{\frac{5}{2}}}{dx} = \frac{5}{2} x^{\frac{5}{2}-1} = \frac{5}{2} x^{\frac{3}{2}}$$

- We have converted 1 to 2 divided by 2 as we were shown in chapter 1.
- Thus the power rule can be expanded to all powers of x, not just integers.

4) Find the derivative of $\dfrac{x^2-1}{x+4}$

- We can't divide out x's because addition and division are not directly related (review fractions in chapter 1)
- Since we can't simplify any further we should just take the derivative.
- We will use the quotient rule, defining our variables.

$$g(x) = x^2 - 1$$

$$h(x) = x + 4$$

- Applying our quotient rule

$$\frac{\dfrac{dg(x)}{dx}h(x) - g(x)\dfrac{dh(x)}{dx}}{\left[h(x)\right]^2} = \frac{2x(x+4) - (x^2-1)(1)}{(x+4)^2}$$

- Cleaning up by simplifying we find

$$\frac{2x(x+4) - (x^2-1)(1)}{(x+4)^2} = \frac{x^2 + 8x + 1}{(x+4)^2}$$

Chapter 4.1 - Integration

Area is a two dimensional metric that measures the amount of space enclosed within a shape. Area for a rectangle is defined as the product of length times the width.

$$\text{Eq 4.1.1} \qquad A = LW$$

This formula can be generalized to a triangle as half the area of a rectangle with comparable length and height. With these two ideas a whole assortment of polygons can be cut up into triangles and rectangles and the area can be found quickly and painlessly. But we aren't going to learn this technique because this technique has a very obvious flaw. Our world is covered with irregular shapes. The shape of leaves, mountains, hills, even our bodies contain shapes that are not perfectly linear. How do we find the exact area covered by say a circle, an ellipse, a bounded parabola? Take for instance the graph below. How do we find the area contained underneath our graph?

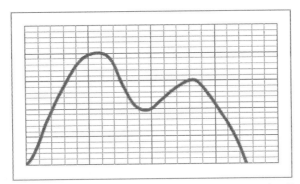

Well, to tell the truth we make a guess, but a very involved guess. The first guess we make is to use the area of a rectangle. We take the largest value of f(x), called the global maximum, and take this as our height. Then we take the entire domain of x that is graphed, given by $x_b - x_a$ and use this as our width, thus our first approximation is a very large rectangle:

$$A \approx f(x)_{max} (\Delta x)$$

We use the squiggly equals sign to denote that it is an approximation and not an exact answer. We have also taken the liberty of reintroducing the delta x to mean the change in x from b to a, just as in the derivative argument. This first approximation is severely overestimated, but one will remember this is only a starting point. To find an approximation of the area under the graph better than this first guess we can separate our function into two partitions as shown below:

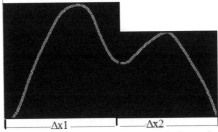

We can interpret two rectangles, one with a width Δx_1 and the other with a width Δx_2. To express the shaded area mathematically we need to find the area of both of these rectangles. Using the area function of a rectangle we can multiply the highest functional value in the first partition (known as the local maximum) by the distance covered by Δx_1. Then we could take the local maximum in the second partition and multiply that by the width Δx_2. Finally, with both areas approximated we will add the two statements together to give the total approximation of the area.

$$A \approx \left[f(x)_{local} \left(\Delta x \right) \right]_1 + \left[f(x)_{local} \left(\Delta x \right) \right]_2$$

It is important to note that the two partitions do not necessarily have to be equal in width or height and as such they cannot be added through the powers of multiplication. Again our approximation of the total area is not exact so we must use the squiggly equals sign. We have, however, guessed a more accurate area below our function than using only one rectangle. If we continue with this process of chopping up our graph into smaller rectangles and adding all of their respective areas (found by the local maximums multiplied by their respective delta x's) then we will, in theory, get better approximations. Below are graphs showing this process for 3 partitions, 6 partitions, and 10 partitions. As we can see, the area covered by the rectangles is starting to model the actual curve. This might not seem exciting now but marvelous things are in store for us.

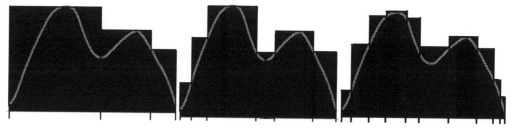

The best way to convey this process mathematically is to use the Riemann sums developed in chapter 2. We will use the form of:

$$A \approx \sum_{L=1}^{n} f\left(x_L \right) \Delta x_L$$

Recall this designates that we will multiply our function evaluated at a given x (our height) by the partition of Δx (our length). This gives us an approximation for the area of that particular rectangular partition. Once we've done that for all of our partitions, which number a total of n, we only need to add all n areas together to get the approximation for the whole area. Notice how the total summation of Δx will equal the total width covered which does not increase or decrease. In essence as the number of partitions increase the size of a particular Δx will decrease, but the new partitions created by definition fills in the lost space. If the total width were to change then we would not be dealing with the same area space and wouldn't be tackling the same problem, which doesn't help us.

Even though our solution seems to model the graph very well, our guess is ultimately still just an approximation and mathematics is an exact art. To advance our argument we borrow a question used in the formulation of the derivative. Notice that our approximation gets better the more rectangles we use in our attempt to guess the total area. What would happen if we kept increasing the total amount of rectangles to a very large number? Naturally the only way to fit more rectangles into the same finite width is for our partition width Δx to become smaller. But how many partitions are considered large? Although a million or a billion might seem like a lot, and it is in our finite human terms, this isn't large enough for math purposes. The only way to mathematically ensure an exact answer to the area under a general function is to increase the amount of rectangles to infinity, which will effectively squeeze the iteration of Δx to zero. Unfortunately, when this happens we get another conundrum. If Δx goes to zero then anything multiplied by zero will be itself zero and the total summation of zero areas would be zero.

$$\sum_{L=1}^{\infty} f(x_1)\Delta x = f(x_1)0 + f(x_2)0 + \cdots + f(x_\infty)0 = 0$$

But that doesn't get us anywhere near the answer. Obviously the area is greater than zero because we can see it! To solve this problem we can take the same trick we used in the derivative argument. We will let Δx approach zero (but never reach it!) which will allow n to approach infinity (but never reach it!). Since Δx never reaches zero, then we still have the assumption that if we add all of our Δx together we would find our total width of the graph. With this argument in hand we can define the area under the graph as:

Eq 4.1.2

$$A = \lim_{n \to \infty} \sum_{L=1}^{n} f(x_L)\Delta x_L = \int_{a}^{b} f(x)dx$$

We pronounce the symbol on the rightmost side as "The integral of f(x) from a to b." Notice that we have used lots of the arguments from the derivative in our formulation of the integral. One will recall that in mathematics there is always a way to manipulate an idea, and then once we've had our fun a way to take the inverse manipulation to return to our original idea. For addition it was subtraction, for multiplication it was division, and for the derivative it is integration.

To explain this let's think about speed. Speed is the distance covered divided by the time it takes to make a journey. We learned in the last chapter that to find our instantaneous speed we simply took the derivative at a specific point. Thus:

Eq 4.1.3 $\qquad s(t) = \dfrac{dp(t)}{dt}$

Since we understand how to get the speed from the time derivative of position, we now contemplate how to find the distance covered from a given speed. The perfect tool for this job is the inverse manipulation, integration.

We can think of it like this. Let's say you travel at a constant speed, 5 miles for every hour in this case. What then is your position in 2 hours? Well its (5 miles per hour)(2 hours) giving 10 miles traveled from your datum point. In three hours it will be 15 miles from your datum point and at 4 hours it will be 20 miles from your datum point. Our process might seem silly but it turns out this is how we evaluate an integral. We take the area under the speed curve from the datum point (t=a) to the point of interest at a later time (t = b) and this gives us our distance covered over an interval t. With our integral notation developed above we can find the area under any type of curve.

To further explore our previous example in the general case we will multiply both sides of our speed equation by (dt)

$$s(t)(dt) = \frac{dp(t)}{dt}(dt)$$

We are allowed to do this because A) dt is not zero, it is infinitesimally close to zero, but not zero. B) the multiplication property of equality allows us to do this. Noticing that dt divided by itself is 1. We find:

$$s(t)(dt) = dp(t)$$

Now we take the integral of both sides and find:

E.Q. 4.14

$$\int_a^b s(t)dt = \int_a^b dp(t)$$

To complete this thought process we need to evaluate the right hand side of the equation. To do this we need to first construct our logical argument. Hopefully we have learned by now that math is not numbers but logical arguments. Recall the definition of the integral as the summation of all the partitions.

$$A = \int_a^b dp(t) = \lim_{n \to \infty} \sum_{L=1}^{n} \Delta p(t)_L$$

Recall that n never reaches infinity and therefore $\Delta p(t)_L$ never reaches zero either. This allows us to add all of the iterations of $\Delta p(t)$ up for the total width given by $p(b) - p(a)$. Another place that our insistence to think in malleable ideas rather than concrete examples pays off. Thus:

$$A = p(b) - p(a)$$

Combining this new knowledge with the original equation we didn't know how to do:

$$\int_a^b s(t)dt = \int_a^b dp(t) = p(b) - p(a)$$

Or written another way

Eq 4.1.5

$$\int_a^b \frac{dp(t)}{dt}dt = [p(t)]_a^b$$

The statement on the right hand side of equation 4.1.5 is the shorthand way to write p(b) – p(a). This is known as the First fundamental theorem of calculus. The first fundamental theorem of calculus states that the integral can be evaluated by taking the difference of the antiderivative evaluated at the bounds. The antiderivative is a function whose derivative produces the function inside of the integral that we now seek to integrate. Although this seems confusing, this antiderivative property of integrals is why evaluating integrals is more of an art than a science. In fact most of the integration rules that we use are based upon the results for differentiation. Fortunately we can still use logic to find a few general statements concerning the integral, which we do in the next section.

Now that we have introduced the first fundamental theorem of calculus it would prove better to forgo example problems and ponder what the integral is. The integral consists of the area between a graph and the x-axis. The derivative on the other hand is the instantaneous change at a point. It is a marvel that these two ideas, area and instantaneous change, could be intertwined and interpreted as opposite sides of a coin. What an interesting idea.

Chapter 4.2 - Advanced Integration

With our understanding of the fundamental theorem of calculus under our belt,

$$\int_a^b \frac{dp(x)}{dx}\,dx = [p(x)]_a^b$$

We will turn now to advanced integration topics. Integration, recall, is the inverse mathematical manipulation of derivation. In fact we burrowed a lot of information used in the derivative argument to advance the argument of integration. This trend will continue and we will use each derivation rule to develop a corresponding integration rule.

Constant Coefficient

Recalling the constant coefficient rule developed in chapter 3:

$$\frac{d[g(x)(c)]}{dx} = c\frac{d[g(x)]}{dx}$$

We will use our 1st fundamental theorem to produce an integration statement.

$$\int_a^b c\frac{dg(x)}{dx}\,dx = \lim_{n\to\infty}\sum_{L=1}^n c\Delta g(x)_L = c\lim_{n\to\infty}\sum_{L=1}^n \Delta g(x)_L = c\int_a^b \frac{dg(x)}{dx}\,dx$$

This argument is a recap of equation 2.4.2. Just as we were able to remove a constant coefficient outside of the derivative, we are able to remove a constant coefficient outside of the integration. Thus our constant coefficient can be moved outside the integral.

Eq 4.2.1

$$\int_a^b c\frac{dg(x)}{dx}\,dx = c\int_a^b \frac{dg(x)}{dx}\,dx$$

Power rule of integration

Recalling the power rule of derivation:

$$\frac{d\left(x^n\right)}{dx} = \lim_{\Delta x\to 0}\frac{(x_0 + \Delta x)^n - \left(x_0\right)^n}{\Delta x} = nx^{n-1}$$

Let's see if we can't find a similar rule for integration:

$$\int_a^b x^n\,dx$$

To do this we will burrow from our Fundamental rules again by combining it with the power rule of derivation:

$$\frac{d(x^n)}{dx} = nx^{n-1}$$

$$\int_a^b \frac{d(x^n)}{dx} dx = \int_a^b nx^{n-1} dx$$

We are allowed to take the integral as long as we do the exact same thing to both sides. Evaluating the integrals:

$$[x^n]_a^b = n \int_a^b x^{n-1} dx$$

Where the first fundamental theorem was used on the left hand side and the constant was pulled out on the right hand side (the rule we just found). Finalizing our answer by dividing both sides by n and then switching the left and right sides and setting the integrated power to n for aesthetic purposes we find the following:

Eq 4.2.2

$$\int_a^b x^n dx = \left[\frac{1}{n+1} x^{n+1} \right]$$

$$n \neq -1$$

But it is important to realize that both equations mean the exact same thing. We will add a power to x and then divide by the new power. Our rule works except for one glaring example. If we take the integral of x^{-1} we would then get $\frac{x^0}{0}$. But we cannot divide by zero! The solution for this derivative of this particular power of x goes beyond the scope of this book. We will define it as the natural log of x (ln(x)) and will let the interested reader approach this through extended reading.

Superposition

Turning our attention to the law of superposition, recall our superposition rule for derivatives:

$$\frac{df(x)}{dx} = \frac{dr(x)}{dx} + \frac{ds(x)}{dx} + ... + \frac{dz(x)}{dx}$$

Taking the integration of both sides we find:

$$\int_a^b \frac{df(x)}{dx} dx = \int_a^b \left[\frac{dr(x)}{dx} + \frac{ds(x)}{dx} + ... + \frac{dz(x)}{dx} \right] dx$$

We wish to look at the right hand side and apply the definition of the integration:

$$\int_a^b \left[\frac{dr(x)}{dx} + \frac{ds(x)}{dx} + ... + \frac{dz(x)}{dx} \right] dx = \lim_{n \to \infty} \sum_{\Delta x=1}^n r(x)\Delta x + s(x)\Delta x + ... + z(x)\Delta x$$

Looking at the right hand side we see a Riemann sum of sums. So we can easily distribute our Riemann sum as described in equation 2.4.3. After this is accomplished we can then return back to our integration notation:

$$\lim_{n\to\infty}\sum_{\Delta x=1}^{n} r(x)\Delta x + s(x)\Delta x + ... + z(x)\Delta x =$$

$$\lim_{n\to\infty}\sum_{\Delta x=1}^{n} r(x)\Delta x + \lim_{n\to\infty}\sum_{\Delta x=1}^{n} s(x)\Delta x + ... + \lim_{n\to\infty}\sum_{\Delta x=1}^{n} z(x)\Delta x$$

Stated formally we have successfully shown superposition to be applicable to integration:

Eq 4.2.3

$$\int_a^b \left[\frac{dr(x)}{dx} + \frac{ds(x)}{dx} + ... + \frac{dz(x)}{dx} \right] dx = \int_a^b \frac{dr(x)}{dx}dx + \int_a^b \frac{ds(x)}{dx}dx + ... + \int_a^b \frac{dz(x)}{dx}dx$$

Integration of a product (known as integration by parts)

Integration will be seen as more of an art than a science. We are tasked with finding the inverse derivative for a given function without knowing what the original function was. To help us we can develop a very powerful tool known as the integration by parts. We look for a way to integrate something that looks like this:

$$\int_a^b \frac{dg(x)}{dx} h(x)dx$$

Where g(x) and h(x) are not related other than having an independent variable of x. Consider the derivative of a product repeated below:

$$\frac{d[g(x)h(x)]}{dx} = \frac{dg(x)}{dx}h(x) + g(x)\frac{dh(x)}{dx}$$

Taking the integral of both sides we find:

$$\int_a^b \frac{d[g(x)h(x)]}{dx}dx = \int_a^b \frac{dg(x)}{dx}h(x)dx + \int_a^b g(x)\frac{dh(x)}{dx}dx$$

Already we see our needed integral jump out at us. The first term on the right hand side is exactly what we are looking for. So to isolate our needed integral we simply subtract the unneeded addition. When we subtract something from one side we must always subtract that same thing from the other side. Also notice that we will use the first fundamental theorem to simplify the integral on the left. Combining all of our changes:

Eq 4.2.4
$$\int_a^b \frac{dg(x)}{dx} h(x)dx = \left[g(x)h(x)\right]_a^b - \int_a^b \frac{dh(x)}{dx} g(x)dx$$

This is known as the integration by parts. As always a numerical example will help clarify after the chapter.

Substitution method

Consider an integral that contains a composite function multiplied by the derivative of the independent function. It will look something like this:

$$\int_a^b f\big(g(x)\big)\left[\frac{dg(x)}{dx}\right]dx$$

Where both $f\big(g(x)\big)$ and $\frac{dg(x)}{dx}$ are taken as functions in their own right. It will take some getting used to see actual problems in this way, but as always a numerical example will help us at the end of the chapter. Assuming we can recognize the second function as the form $\frac{dg(x)}{dx}$ then we can eliminate the dx infinitesimals giving us:

$$\int_{g(a)}^{g(b)} f(g(x))\,dg(x)$$

All we have done is swapped which variable we are integrating with. Note a very important change in the bounds of the integration, where a and b usually sit. Since we have changed the variable of the integration we must also take into account that the bounds change as well. This goes beyond the scope of the book but I whole heartedly suggest reading about it. To complete the analogy we must equate f(g(x)) to the first derivative of a new function P(g(x)). Assuming this is allowable, and assuming integrating is even possible we can write:

$$\int_{g(a)}^{g(b)} f\big(g(x)\big)\,dg(x) = \int_{g(a)}^{g(b)} \frac{dP\big(g(x)\big)}{dg(x)}\,dg(x) = \big[P(g(x))\big]_{g(a)}^{g(b)}$$

We are allowed to take the derivative with respect to any independent variable, as our argument in equation 3.3.6 showed. So with the assumption that f(g(x)) is the derivative of P(g(x)) we can use the fundamental theorem of calculus to solve the integration. IF all of these steps are possible, and most of the time they are not, then we have substituted a very easy form for a difficult form and achieved our solution.

Eq 4.2.5

$$\int_a^b f\big(g(x)\big)\left[\frac{dg(x)}{dx}\right]dx = \big[P(g(x))\big]_{g(a)}^{g(b)}$$

Remember we are substituting the value of a and b for x, which is a part of g(x)! It is worth mentioning again that the observant reader might have recognized that many assumptions were taken to advance the theory of the substitution method. This is because the substitution method only works for very particular circumstances and is not useful in all cases. But the substitution method is a very powerful tool for those specific instances that it does actually work. As always numerical examples follow.

Numerical examples:

1) Take the integral of $x^2 + 3x - 5$ from x = 0 to 8
   - First we use the law of superposition to break apart our terms and integrate each separately.

   $$\int_0^8 x^2 dx + \int_0^8 3x dx + \int_0^8 (-5) dx$$

   - We now pull out the constant coefficient

   $$\int_0^8 x^2 dx + \int_0^8 3x dx + \int_0^8 (-5) dx =$$

   $$\int_0^8 x^2 dx + 3\int_0^8 x dx - 5\int_0^8 dx$$

   - We proceed into the power rule for integration.

   $$\left[\frac{1}{3}x^3\right]_0^8 + 3\left[\frac{1}{2}x^2\right]_0^8 - 5[x]_0^8$$

   - Completing our answer by evaluating

   $$\left[\frac{1}{3}x^3\right]_0^8 + 3\left[\frac{1}{2}x^2\right]_0^8 - 5[x]_0^8 =$$

   $$\left[\frac{1}{3}(8^3 - 0)\right] + 3\left[\frac{1}{2}(8^2 - 0)\right] - 5[8 - 0] =$$

   $$\frac{512}{3} + \frac{192}{2} - 40 \approx 226.67$$

   - Notice how closely our integration process followed the derivative process.

2) Integrate $\displaystyle\int_0^x (x^2 - 1)^{34} 2x dx$

   - We immediately notice that the derivative inside the parenthesis $(x^2 - 1)$ is present.
   - This prompts a hope that we can use the integration chain rule.
   - Defining our functions

   $$g(x) = (x^2 - 1)$$

   - There is an opportunity here, because if the derivative of g(x) is 2x then we would have a very powerful piece of information
   - Crossing our fingers we take the derivative of g(x)

$$\frac{dg(x)}{dx} = 2x$$

- Very exciting as we sense we are getting closer
- We will now simply define our composite function as the missing piece

$$f(g(x)) = (g(x))^{34} = (x^2 - 1)^{34}$$

- In essence we are trying to produce a gateway from our current form into an easier form
- Substituting our definitions

$$\int_0^x (x^2 - 1)^{34} \, 2x \, dx = \int_0^x (g(x))^{34} \frac{dg(x)}{dx} \, dx = \int_0^x (g(x))^{34} \, dg(x)$$

- Hence we use the power rule with g(x) in mind

$$\int_0^x (g(x))^{34} \, dg(x) = \left[\frac{1}{35} (g(x))^{35} \right]_{g(0)}^{g(x)}$$

- Making one last substitution we find:

$$\left[\frac{1}{35} \left(g(x)^{35} - g(0)^{35} \right) \right] = \frac{1}{35} \left((x^2 - 1)^{35} - (0^2 - 1)^{35} \right) = \frac{1}{35} (x^2 - 1)^{35} + \frac{1}{35}$$

- Long but not difficult

3) Integrate $\displaystyle\int_0^3 x(x^2 - 4)^2 \, dx$

- We again see the possibility of a chain rule. We use the substitution

$$g(x) = x^2 - 4$$

$$\frac{dg(x)}{dx} = 2x$$

- Notice however that we don't have 2x in the original question, we only have x.
- We can get around this by multiplying by a funny way of saying 1.

$$\int_0^3 x(x^2 - 4)^2 \, dx = \int_0^3 \frac{2}{2} x(x^2 - 4)^2 \, dx = \frac{1}{2} \int_0^3 (x^2 - 4)^2 \, 2x \, dx$$

- Now we have the 2x that we needed so we could insert our substitution
- Substituting

$$\frac{1}{2} \int_{g(0)}^{g(3)} (g(x))^2 \frac{d(g(x))}{dx} \, dx = \frac{1}{2} \int_{g(0)}^{g(3)} (g(x))^2 \, dg(x)$$

- Now we just take the integral power rule and don't forget the half outside the integral

$$\frac{1}{2}\int_{g(0)}^{g(3)} \left(g(x)\right)^2 dg(x) = \frac{1}{2}\left[\frac{1}{3}\left(g(x)\right)^3\right]_{g(0)}^{g(3)}$$

- Making our final substitution and evaluating we find:

$$\frac{1}{2}\left[\frac{1}{3}\left[\left(3^2-4\right)^3-\left(0^2-4\right)^3\right]\right] = \frac{189}{6} = 31.5$$

4) Integrate $\displaystyle\int_0^2 \left(x^2+3\right)\left(x^3\right)dx$

- We will use integration by parts
- First we define our functions

$$\frac{dg(x)}{dx} = x^3$$

$$h(x) = x^2 + 3$$

- Taking the integral and the derivative respectively.

$$g(x) = \frac{1}{4}x^4$$

$$\frac{dh(x)}{dx} = 2x$$

- Now we have all four functions needed to use the integration by parts.

$$\int_0^2 x^3\left(x^2+3\right)dx = \frac{1}{4}x^4\left(x^2+3\right) - \int_0^2 \frac{1}{2}x^5 dx$$

- Where we simplified the integral on the right. We now need to take the integral on the right to further simplify the answer.

$$\int_0^2 x^3\left(x^2+3\right)dx = \left[\frac{1}{4}x^4\left(x^2+3\right)\right]_0^2 - \frac{1}{12}\left[x^6\right]_0^2$$

- Evaluating

$$\left[\frac{1}{4}2^4\left(2^2+3\right) - \frac{1}{4}0^4\left(0^2+3\right)\right] - \frac{1}{12}\left[2^6 - 0^6\right] \approx 22.667$$

- Notice that the answer could be found by distributing the x^3 then integrating.

The Calculus Direct

Chapter 5: Application of the Calculus

We have been through a lot of work, and I promised you great things would come if we obtained a basic understanding of calculus. Now I will deliver.

Application 1: Position, velocity, acceleration

Why do things fall down? How does gravity work? What is gravity? Unfortunately none of these questions have been answered fully. However we can use our knowledge of calculus to figure out how things act under gravity.

We've already shown that velocity is the derivative of distance with respect to time:

$$\frac{dp(t)}{dt} = v(t)$$

Taking a further derivative we find the definition of acceleration.

$$\frac{dv(t)}{dt} = a(t)$$

This shows that the second derivative of position is acceleration:

$$\frac{d^2(p(t))}{dt^2} = \frac{dv(t)}{dt} = at$$

We can also go the other way, namely the integral of acceleration is velocity and the integral of velocity is position.

$$\int_a^b a(t)dt = v(t)$$

$$\int_a^b v(t)dt = p(t)$$

Application 2: Newton's Second Law

We've all heard that force equals mass times acceleration since we were children. Well, it turns out we have been lying to you since you were children. I'm not saying it's wrong, but I think it should stop. The actual formulation of Newton's law is

$$F = \frac{d(momentum)}{dt}$$

Momentum is the mass of an object multiplied by the velocity of the object.

$$F = \frac{d(momentum)}{dt} = \frac{d(mv)}{dt}$$

We can quickly see the derivative of a product, applying our rule:

$$F = \frac{d(mv)}{dt} = \frac{dm}{dt}v_0 + m_0\frac{dv}{dt}$$

This equation is the basis for rockets! As rockets burn fuel they accelerate upwards, however as they burn the fuel they become lighter and lighter. You must know this equation to get to the moon, and now you understand the basic principles. But we've already been to the moon, we'd like to go to mars now. You now understand how to get us there, will you dare? (Ok. There are a few more equations to get there, but this is the first step)

Application 3: 1st law of Thermodynamics

Thermodynamics is the governing principles of your internal combustion engine that takes you to work, school, across the world in jets, or to the moon in rockets. It is one of the building blocks of our modern society. Its 1st law states that the energy in a closed system is equal to the work input plus the heat input.

$$E = W + Q$$

It also states that during a cycle the change in energy is zero. A cycle is a set of processes that returns the system to its original state. In your car we move through an explosion process, an exhaust process, and intake process, and finally we're ready for another explosion process. Your radiator is in place to keep your engine block at a steady temperature so that everything is the same for the next explosion process thus returning the system to the starting state. Since we have returned the car to the starting state then our energy has returned to its datum point. To explore interesting characteristics of the 1st law we will take the derivative of the 1st law of thermodynamics with respect to a cycle. Notice that Energy hasn't changed so the derivative is 0

$$0 = \frac{d(W)}{d(cycle)} + \frac{d(Q)}{d(cycle)}$$

Where we have used the superposition property to break up the sum. Also notice that we took the derivative with an idea (cycle) This is ok, because we have learned to think in ideas and not numbers. Now the pay off, we subtract the first term from both sides:

$$-\frac{d(W)}{d(cycle)} = \frac{d(Q)}{d(cycle)}$$

What this means is that if we input heat into a system we get work out of the system. This is what your car does, we create heat by setting a tiny fire in your car and work is pushed out of the system by the piston moving up and down ultimately turning your tire.

What's cooler (pun intended) is if we put work into the system, then we pull heat out of the system and the object cools down. We have just invented an air conditioner. Well, in theory of course. We'd also have to enter material conduction properties, fluid flow properties, latent heat of evaporation, pressure-temperature phase change, and so on. (note: this 1st law assumes a reversible process, for more on that I suggest reading a thermodynamics book)

The Calculus Direct

Application 4: mass of a conglomerate

A conglomerate is a collection of different materials. We define density as mass divided by volume. Therefore mass is:

$$m = \rho V$$

Where rho stands for density. Let's consider a large collection of particles, they may be exactly the same, they may not be, we just want to count all of them up. We do this by making partitions of the volume, multiplying by the local density, and then adding up all the groups. This sounds a lot like integration. Running with this idea we proposition:

$$m = \int_V \rho\, dv$$

The integration is taken over all of the volume. Notice that now we don't care what shape the object is, we can simply take the integral and find our absolute amount. We can use this formula as a starting point to find center of mass, material mass, moment of inertia, density of heavenly bodies, and so on.

Application 5: Fluid Dynamics

A fluid is defined as a substance that flows readily under a shear stress. We can measure how readily a fluid responds to a shear stress by its viscosity, which is a material property. The formula for the flow of a fluid is

$$\tau = \mu \frac{du(h)}{dh}$$

Where tau is the shear stress, mew is the viscosity and u(h) is the speed of the fluid at a height h.

Application 6: Heat Transfer

Everyone's computer, car, and even their bodies create heat through the course of work, we've already seen how the car creates work (by an input of heat) But how is heat transferred? One way is by conduction. The heat transferred by conduction per area is given by:

$$\frac{dQ}{dA} = -k \frac{dT(x)}{dx}$$

Q is heat, A is area, k is a material property, and T(x) is temperature at a given x. We can find the total heat transfer of a surface by taking the integral with respect to area:

$$\int_A \frac{dQ}{dA} dA = \int_A -k \frac{dT(x)}{dx} dA$$

$$Q = \int_A -k \frac{dT(x)}{dx} dA$$

Notice how all of the terms inside the second integral are not area dependent, therefore they are constant as far as area is concerned:

$$Q = \int_A -k\frac{dT(x)}{dx}\,dA = -kA\frac{dT(x)}{dx}$$

And we have found the conduction law.

Application 7: Marginal Cost

In business, marginal cost is the cost of producing one more unit of goods. The question is how to find the most efficient use of resources at your disposal given a finite amount of capital. This is a very practical example of calculus because everyone has finite capital, some just more finite than others. To find the marginal cost of an amount of units we use the formula

$$MC(q) = \frac{d(FC + VC(q))}{dq}$$

Which states that the derivative of the sum of the variable cost (VC(q)) and the fixed cost (FC) with respect to the quantity (q) is the marginal cost at that particular quantity q.

Since the fixed costs are by definition fixed (not variable) we can use the superposition property to break up this derivative into two derivatives.

$$MC(q) = \frac{d(FC)}{dq} + \frac{dVC(q)}{dq} = \frac{dVC(q)}{dq}$$

Notice that the derivative of the first term goes to zero because a fixed cost never changes. A typical use of this equation would be to find the optimum quantity of items a facility should produce. I did not go over this use of the calculus in this book. The interested student should explore higher level calculus books for more on optimization. For instance one could conceivable optimize any assortment of variables to produce the mathematically best way to do some endeavor, at least under the current assumptions that the endeavor is taken with. Without stalling any more I will present how to find the optimum quantity. The idea is to take the derivative of the marginal cost and set that derivative equal to zero. As for why this is... I suggest reading a physics book on ballistics.

$$\frac{dMC(q)}{dq} = 0 = \frac{d^2VC(q)}{dq^2}$$

Closing statement:

As said before, this book is a beginning, not an end all to mathematical understanding. You have successfully completed a basic understanding of the ideas that have shaped our modern world. With this equation you can now understand and even add to the technical analytical world that governs our lives. You have earned a great power. It is up to you to choose what you will do with it. Will you invent a new process that makes life better for the people who inhabit this earth? Will you embark on a great journey to new places and explore where no human has been before? Other souls have created all of the wonders that surround us with the ideas of calculus presented in this book... what will you do?

14229189R00056

Made in the USA
Lexington, KY
15 March 2012